CONSERVATION
MEDICINE

A MULTIDISCIPLINARY APPROACH TO
MAINTAINING GLOBAL BIODIVERSITY

DALE J. BLOCK

ISBN: 979-8-89633-057-8 (sc)
ISBN: 979-8-89633-058-5 (e)

PAGE
SOLUTIONS
Page Solutions
541 Buttermilk Pike
Crescent Springs, KY 41017

Printed in the United States of America

CONTENTS

PREFACE

The health and resilience of our planet are inseparably tied to its inhabitants and ecosystems. As we face an era of unprecedented environmental challenges, the intricate connections between biodiversity conservation, human-induced climate change, and global healthcare delivery demand urgent attention. Rapid species extinction, habitat destruction, and climate instability have pushed the Earth toward a tipping point, threatening the delicate balance of life. *Conservation Medicine: A Multidisciplinary Approach to Maintaining Global Biodiversity* offers a vital framework for understanding and addressing the intersection of ecological integrity, health systems, and clinical science. Serving as both a warning and a call to action, this book underscores the urgency of coordinated efforts to prevent further ecological collapse.

This book unites a diverse array of interdisciplinary and multisectoral subject matter spanning veterinary medicine, human medicine, public health science, ecology, environmental science, socioeconomics, and public policy. By bringing together these disciplines, we illuminate the necessity of a holistic, comprehensive, systems-based approach to *Conservation Medicine*—one that not only recognizes the interdependence of all living organisms but also fosters innovative solutions to mitigate emerging threats. Through this integration, we aim to highlight *Conservation Medicine's* role in protecting all species, curbing the spread of infectious diseases, and preserving the resilience of ecosystems that sustain life. The book explores novel strategies, and innovative multisectoral collaborations, to demonstrate how interdisciplinary cooperation and communication can drive meaningful conservation outcomes.

The urgency of this work cannot be overstated. Habitat destruction, deforestation, climate instability, and the exploitation of natural resources have accelerated global biodiversity loss at an alarming rate. These environmental disruptions have direct and devastating consequences for the public's health, as demonstrated by the rise of zoonotic diseases such as COVID-19, Ebola, and avian influenza. The complex interactions between wildlife, domestic animals, and human populations create pathways for disease spillover, underscoring the need for proactive, empirically-driven, and evidence-based interventions. Without decisive action, we risk entering an era where pandemics and natural disasters become more frequent, food security is further compromised, and entire ecosystems collapse under the weight of human impact. *Conservation Medicine: A Multidisciplinary Approach to Maintaining Global Biodiversity* provides a critical lens through which we can anticipate and address these challenges, emphasizing proaction over reaction. By advancing research in emerging infectious diseases, advocating for habitat protection, and integrating systems of health and environmental policies, *Conservation Medicine* offers a pathway to resilience and sustainability.

Beyond disease ecology, this book delves into the broader implications of biodiversity loss, including its effects on food security, water quality, climate regulation, and even global economic stability. Ecosystems' disruptions directly affect ecosystem services that all living things rely on, from pollination and clean air to carbon sequestration and soil fertility. The consequences of failing to protect biodiversity extend beyond wildlife extinctions—they threaten the fundamental resources that sustain human civilization. Through carefully chosen content that highlights an interdisciplinary and multisectoral approach to global biodiversity conservation, the primary goal of this book is to provide a roadmap for leaders, researchers, healers, and policymakers committed to shaping a sustainable future on Earth for all living things. The interdisciplinary and multisectoral approach presented in these pages underscores that human health cannot be disentangled from the health of our planet. More importantly, it presents practical solutions for reversing ecological and ecosystem damage, promoting sustainable development, and ensuring equitable access to natural resources for today and tomorrow.

This book serves as both an educational resource and a call to action for students, researchers, practitioners, and decision-makers. It aims to inspire a new generation of *Conservation Medicine* professionals to champion integrated systems approaches to planetary health. Readers are encouraged to view *Conservation Medicine* not as a specialized niche but as an essential pillar of global healthcare systems, driving long-term environmental sustainability.

The solutions presented herein, require commitment at all socio-ecological levels—from grassroots community initiatives to international public policy changes—and must be implemented with urgency, tenacity, and perseverance. By working across disciplines and geographies, we can drive meaningful change—preserving biodiversity, strengthening the public's health, and ensuring a balanced coexistence between humans and the natural world.

You are invited to engage with the ideas and strategies presented in these pages, to challenge conventional societal thinking, and to contribute to a growing movement dedicated to safeguarding the intricate web of life that sustains us all. The future of our planet hinges on the choices we make today, and through the lens of *Conservation Medicine*, we have the opportunity to forge a healthier, more sustainable world for all species, now and for future generations to come.

Dale J Block, MD, MBA

Spring 2025

1.0

Introduction

1.1 Global Biodiversity Threats and the Need for a Multidisciplinary Response.

1.1.1 Introduction.

Biodiversity—the vast array of life on Earth—is essential for ecosystem stability and the health, well-being, and resilience of all living organisms. Ecosystems provide critical services, including food security, climate regulation, water purification, and cultural enrichment. However, human activities pose severe and immediate threats to biodiversity, demanding urgent, multidisciplinary action to address this crisis. This discussion will examine the primary global threats to biodiversity and emphasize the need for integrated strategies spanning scientific, economic, political, and social domains.

1.1.2 Major Threats to Global Biodiversity.

1. *Habitat Destruction and Land-Use Change.*

   The expansion of agricultural lands, urbanization, and infrastructure development have significantly reduced natural habitats, leading to species decline and ecosystem fragmentation (Foley et al., 2011). Deforestation, particularly in tropical rainforests, has had devastating consequences for biodiversity, with approximately 80% of terrestrial species residing in forested eco-systems (Gibson et al., 2011).

2. *Anthropocene Climate Change.*
   Anthropocene (i.e., human activity as the most dominant influ-ence) climate change exacerbates biodiversity loss by altering spe-cies distributions, disrupting migration patterns, and increasing extreme weather events (Parmesan & Yohe, 2003). Ocean acid-ification, caused by increased CO_2 absorption, has threatened marine biodiversity, particularly coral reefs, which support 25% of marine species (Hoegh-Guldberg et al., 2007).

3. *Pollution.*
   Pollution, including plastic waste, heavy metals, and agricultural runoff, has detrimental effects on ecosystems. Marine pollution, particularly plastic debris, affects more than 700 marine spe-cies through ingestion and entanglement (Wilcox et al., 2015). Pesticides and industrial waste contribute to soil and water con-tamination, further threatening biodiversity.

4. *Overexploitation of Natural Resources.*
   Overfishing, poaching, and unsustainable logging have led to significant declines in wildlife populations. The global demand for seafood has resulted in the overexploitation of 34% of fish stocks, pushing several species toward extinction (FAO, 2020). Similarly, the illegal wildlife trade has endangered species such as pangolins, rhinos, and elephants (Rosen & Smith, 2010).

5. *Invasive Species.*
   Human activity has facilitated the spread of invasive species, which outcompete native species, disrupt food webs, and alter habitats (Simberloff et al., 2013). Invasive species are responsible

for approximately 40% of known species extinctions (Bellard et al., 2016).

1.1.3 The Need for a Multidisciplinary Response.

Given the complexity of biodiversity threats, addressing them requires collaboration across multiple disciplines.

1. *Scientific and Technological Innovations.*
   Advancements in ecological research, biotechnology, and remote sensing are crucial for biodiversity conservation. DNA barcoding helps identify species at risk (Hebert et al., 2003), while satellite imagery provides real-time data on habitat changes (Hansen et al., 2013).

2. *Economic and Policy Interventions.*
   Market-based mechanisms, such as payments for ecosystem services (PES) and sustainable certification programs, incentivize biodiversity conservation (Wunder, 2007). Stronger international agreements, such as the Convention on Biological Diversity (CBD), are essential for coordinated global action (CBD, 2020).

3. *Community Engagement and Indigenous Knowledge.*
   Local communities play a vital role in biodiversity conservation efforts. Integrating traditional ecological information and knowledge and expertise with modern conservation strategies has proven effective in managing biodiversity (Berkes et al., 2000).

4. *Interdisciplinary and Multisectoral Collaboration.*
   Effective biodiversity conservation requires cooperation between ecologists, policymakers, economists, and social scientists. A holistic, integrated and comprehensive approach ensures sustainable empirically-driven evidence-based management of eco-

systems while balancing economic and social needs of society (Rockström et al., 2009).

1.1.4 Conclusion.

Biodiversity loss is an urgent global crisis with profound ecological, economic, political, and social implications. The rapid decline of species, ecosystems, and genetic diversity is driven by habitat destruction, man-made climate change, pollution, overexploitation of natural resources, and unchecked invasive species proliferation. These threats not only disrupt ecosystems and the services they provide but also jeopardize food security, human health, and economic stability. Addressing this complex challenge requires a holistic, integrated, comprehensive, and multidisciplinary response that integrates scientific research, education, and practice, international and regional public policy development, aligned economic incentives, and active local community engagement and empowerment.

Scientific advancements can enhance biodiversity conservation efforts through habitat restoration, species monitoring, and innovative biotechnology solutions. Policymakers must implement robust environmental laws and regulations, enforce sustainable land-use practices, and promote international cooperation to protect the degradation of biodiversity hotspots. Economic strategies, such as sustainable resource management, green investments, and incentives for biodiversity-friendly industries, can drive long-term conservation success. Additionally, empowering local communities through education, traditional knowledge integration, and participatory decision-making is vital for fostering environmental stewardship.

By uniting these diverse disciplines, humanity can develop resilient, adaptive solutions to mitigate biodiversity loss and safeguard Earth's natural heritage today and tomorrow.

1.2 Integrating Human, Animal, and Environmental Health to Mitigate Biodiversity Loss.

1.2.1 Introduction.

As previously discussed, biodiversity loss is a pressing global challenge with profound implications for human health, animal populations, and the stability of ecosystems. The inextricable interdependence between humans, animals, and the environment necessitates a holistic, integrated, and comprehensive approach to biodiversity conservation and public health practice. The *One Health* framework, which recognizes the inextricable interconnectedness of human, animal, and environmental health, provides a holistic and integrated strategy for addressing biodiversity loss while promoting sustainable development (Destoumieux-Garzón et al., 2018). This discussion explores the importance of integrating human, animal, and environmental health to mitigate biodiversity loss, emphasizing the roles of disease transmission, habitat destruction, man-made climate change, and policy-driven biodiversity conservation efforts.

1.2.2 The Interconnection of Human, Animal, and Environmental Health.

The health of humans, animals, and ecosystems is inextricably interconnected. Human activities, such as deforestation, industrial agriculture, and urban expansion, have disrupted natural habitats, leading to biodiversity loss and increased human-wildlife interactions. These interactions have heightened the risk of zoonotic disease transmission, as evidenced by outbreaks such as SARS-CoV-2 (i.e., COVID-19), Ebola, and avian influenza (Daszak et al., 2020). Habitat destruction not only threatens wildlife but also affects ecosystem services that humans rely on, such as pollination, clean water, and climate regulation (Díaz et al., 2019).

Similarly, declining biodiversity weakens ecosystems' resilience to environmental stressors, exacerbating climate change impacts. For example, deforestation contributes to greenhouse gas (GHG) emissions, while the loss of keystone species can lead to cascading ecological consequences, disrupting food chains and altering habitat structures (Ceballos et al., 2017). Addressing these challenges requires a coordinated approach that integrates human, animal, and environmental health considerations.

1.2.3 Mitigating Biodiversity Loss.

1. *Preventing Zoonotic Disease Transmission.*
   One of the most urgent reasons for integrating human, animal, and environmental health is to prevent zoonotic disease outbreaks. Approximately 60% of emerging infectious diseases originate from animals, and biodiversity loss has been linked to increased pathogen spillover (Jones et al., 2008). By preserving natural habitats and maintaining healthy wildlife populations, the risk of zoonotic transmission can be reduced.

   Moreover, sustainable agricultural and livestock practices can help prevent the spread of diseases between domesticated animals and wildlife. Policies that regulate wildlife trade and enforce biosecurity measures in farming can further minimize the risk of zoonotic diseases crossing species barriers (Karesh et al., 2012).

2. *Sustainable Land Use and Habitat Protection.*
   Land-use change is one of the primary drivers of biodiversity loss. Urban expansion, deforestation, and intensive agriculture have led to habitat fragmentation and species extinction. Integrating environmental health science into land management policies can help protect critical ecosystems. Strategies such as reforestation, the establishment of protected areas, and sustain-

able agricultural practices can preserve biodiversity while supporting local economies (Chaudhary & Brooks, 2018).

Additionally, implementing nature-based solutions (NbS)—such as wetland restoration and agroforestry—can enhance ecosystem resilience and mitigate the impacts of man-made climate change. These approaches benefit both biodiversity conservation and human and animal health and resiliency by improving water quality, enhancing carbon sequestration, and providing sustainable livelihoods for communities that depend on natural resources (Seddon et al., 2020).

3.   *Anthropocene Climate Change Mitigation and Adaptation.*
Anthropocene climate change and biodiversity loss are interlinked crises that require an integrated response. Rising temperatures, extreme weather events, and shifting precipitation patterns have disrupted ecosystems, altering species distributions and threatening biodiversity (Scheffers et al., 2016). A holistic, integrated, and comprehensive approach promotes adaptive conservation strategies, such as assisted migration of species, ecosystem-based adaptation, and climate-smart agriculture.

Reducing greenhouse gas (GHG) emissions through sustainable energy policies, afforestation, and conservation of carbon-rich ecosystems like peatlands and mangroves can mitigate biodiversity loss while benefiting human and animal vigor. Additionally, improving climate resilience in rural and indigenous communities is essential, as these populations are often the most affected by environmental degradation and biodiversity loss (Reid et al., 2019).

4.   *Strengthening Policy and Global Collaboration.*
Effective biodiversity conservation requires coordinated efforts across disciplines, sectors, and nations. Policymakers must integrate systems of health (e.g., integrated person-centered primary health care, (IPC-PHC)) and environmental considerations into

legislation and regulation, ensuring that economic development does not come at the cost of biodiversity loss. International agreements, such as the *Convention on Biological Diversity (CBD)* and the *One Health High-Level Expert Panel (OHHLEP)*, emphasize the importance of cross-sectoral collaboration in addressing biodiversity loss (WHO, FAO, OIE, & UNEP, 2021).

Investing in interdisciplinary research and education can also drive policy innovation. Strengthening partnerships between scientists, conservationists, public health professionals, and indigenous communities ensures that conservation efforts are evidence-based, culturally inclusive, and sustainable in the long term (Ostrom et al., 1999).

1.2.4 Conclusion.

The interconnectedness of human, animal, and environmental health is fundamental to mitigating biodiversity loss and securing a sustainable future. A holistic, integrated, and comprehensive approach acknowledges the interdependence of ecosystems, emphasizing adaptive strategies that support both biodiversity conservation and public health. Recognizing that ecosystem degradation directly contributes to the emergence of zoonotic diseases, climate instability, and resource scarcity, proactive measures are essential to breaking this cycle of harm.

Preventing zoonotic disease transmission, adopting sustainable land-use practices, mitigating human-induced climate change, and fostering global collaboration are crucial steps toward reversing biodiversity loss. Sustainable agricultural practices, responsible wildlife trade regulations, and habitat restoration efforts must be prioritized to protect ecosystems while maintaining economic and social stability. Additionally, investments in renewable energy, pollution reduction, and climate resilience strategies can create synergies that benefit both environmental and human health.

Strengthening interdisciplinary and multisectoral collaboration across scientific, economic, and policy domains will be critical in safeguarding the planet's natural heritage for future generations. By integrating empirically-derived evidence-based policies with local and indigenous information and knowledge systems, societies can develop context-specific solutions that reinforce ecological resilience. A long-term commitment to education, innovation, and international cooperation will be necessary to ensure that biodiversity conservation remains at the forefront of global health and environmental sustainability initiatives.

1.3 Introduction to Conservation Medicine.

1.3.1 Definition of Conservation Medicine.

Conservation Medicine is an interdisciplinary and multisectoral discipline that examines the complex relationships between human health, animal health, and ecosystem health, with a focus on how environmental changes and biodiversity loss impact disease emergence and overall ecological stability (Aguirre et al., 2002). This field of study integrates information and knowledge from veterinary medicine, human medicine, ecology, public health science, and conservation biology to address global and regional health challenges related to habitat destruction, man-made climate change, pollution, and the spread of zoonotic diseases (Daszak et al., 2004).

Conservation Medicine emerged as a recognized health-related scientific discipline in response to the increasing recognition that human activities—such as deforestation, industrial agriculture, and wildlife trade—are driving biodiversity loss and facilitating the transmission of infectious diseases across species (Travis et al., 2011). Through a holistic, integrated, and comprehensive approach, *Conservation Medicine* seeks to develop foundational expertise, enhance capacity, and implement sustainable strategies that safeguard both biological diversity and human health. Recognizing that the well-being of any single component within an ecosystem is inherently and inseparably linked to the health of the whole, this approach fosters resilience by addressing the complex interdependencies between environmental, animal, and human health (Osofsky et al., 2005).

1.3.2 Key Principles of Conservation Medicine.

1.3.2.1 One Health Approach.

One Health is a collaborative, transdisciplinary, holistic, and integrative approach that recognizes the inextricable interconnectedness of human, animal, and environmental health. It emphasizes the need for cooperation across multiple sectors—including human medicine, veterinary science, environmental science, public health, agriculture, and policy-making—to effectively address health challenges at the human-animal-environment interface (Destoumieux-Garzón et al., 2018; Mackenzie & Jeggo, 2019).

Core Principles of *One Health* include the following:

1. *Interconnected Health Systems.*
   The health of humans, animals, and ecosystems is deeply intertwined. Environmental degradation, biodiversity loss, and man-made climate change directly impact the emergence and spread of infectious diseases, food security, and overall well-being (Gibbs, 2014). For instance, deforestation and habitat destruction increase human-wildlife interactions, facilitating zoonotic disease transmission, as seen in outbreaks of Ebola, SARS, and COVID-19 (Jones et al., 2008).

2. *Zoonotic Disease Prevention and Control.*
   More than 60% of emerging infectious diseases are zoonotic, meaning they originate in animals before jumping to humans (Taylor et al., 2001). *One Health* approaches integrate veterinary and human medicine with environmental science to enhance dis-

ease surveillance, improve early detection, and implement coordinated responses to emerging threats (Gebreyes et al., 2014).

3. *Antimicrobial Resistance (AMR) Mitigation.*
    Overuse and misuse of antibiotics in human medicine, livestock production, and agriculture contribute to the exponential rise of antimicrobial resistance world-wide. *One Health* strategies promote responsible antibiotic stewardship across sectors to preserve the effectiveness of existing treatments and reduce the spread of resistant pathogens (O'Neill, 2016).

4. *Food Safety and Security.*
    Sustainable food production is critical for global health, wellness, and well-being. *One Health* principles advocate for responsible farming practices, reduction of pesticide and antibiotic overuse, and policies that protect ecosystems while ensuring sufficient and safe food supplies (FAO, 2021).

5. *Man-made Climate Change and Environmental Health.*
    Man-made climate change exacerbates health risks by altering disease transmission patterns, increasing the frequency of extreme weather events, and threatening biodiversity conservation. *One Health* emphasizes the need for ecosystem conservation, climate adaptation and mitigation strategies, and ecosystem restoration to support long-term planetary health (IPCC, 2022).

6. *Policy and Global Collaboration.*
    Effective implementation of *One Health* requires collaboration across disciplines, sectors, governments, and international organizations. The World Health Organization (WHO), Food and Agriculture Organization (FAO), World Organisation for Animal Health (WOAH, formerly OIE), and United Nations Environment Programme (UNEP) are key stakeholders in

advancing *One Health* strategies at global and national levels (WHO, 2021).

7. *Applications of One Health*
   a) *Pandemic Preparedness:* Integrated disease surveillance and response systems for zoonotic diseases help with population health management by detecting and containing outbreaks before they escalate.
   b) *Wildlife Conservation:* Protecting biodiversity conservation reduces the risk of pathogen spillover from animals to humans.
   c) *Urban Planning & Health:* Designing cities with green spaces, clean water, and sustainable infrastructure enhances public and environmental health.
   d) *One Health Education & Research:* Expanding interdisciplinary training for medical, veterinary, and environmental professionals fosters innovation and better problem-solving.

By embracing a *One Health* approach, societies can build resilient healthcare delivery systems, mitigate global existential health threats, and promote sustainable development, ensuring the well-being of all life on Earth.

1.3.2.2 Ecosystem-Based Disease Management.

Ecosystem-based disease management is a core principle of *Conservation Medicine*, recognizing that habitat destruction, man-made climate change, and environmental degradation significantly alter disease dynamics and increase health risks for humans, wildlife, and domestic animals. As human activities continue to disrupt natural ecosystems—through deforestation, urbanization, intensive agriculture, and pollution—pathogen transmission patterns shift, often leading to the emergence and re-emergence of infectious diseases (Murray & Daszak, 2013).

Conservation Medicine addresses these challenges by advocating for habitat protection, restoration, and sustainable land-use practices as strategies to mitigate disease risks. Maintaining intact and biodiverse ecosystems can serve as a natural barrier against the spread of pathogens. For example, research on the *dilution effect* suggests that higher biodiversity conservation can reduce disease transmission by decreasing the prevalence of pathogen reservoirs among wildlife populations (Keesing et al., 2010). Conversely, habitat fragmentation and biodiversity loss can lead to an increased concentration of disease-carrying species, facilitating the spillover of zoonotic pathogens to humans and livestock.

Man-made climate change further exacerbates disease risks by altering temperature and precipitation patterns, shifting the geographic range of vectors such as mosquitoes and ticks, and creating favorable conditions for pathogens to thrive. For instance, the spread of vector-borne diseases like malaria, Lyme disease, and dengue fever has been linked to climate-induced habitat changes (Scheffers et al., 2016). *Conservation Medicine* integrates climate adaptation strategies, such as wetland restoration, reforestation, and sustainable water management, to mitigate these risks and enhance ecosystem resilience.

Ecosystem-based disease management also promotes interdisciplinary and multisectoral collaboration between ecologists, public health officials, veterinarians, and policymakers. By integrating ecosystem education, research, and practice with disease surveillance and policy initiatives, *Conservation Medicine* supports proactive health strategies that address the direct and indirect environmental perturbations rather than just treating disease outbreaks after they occur. Efforts such as rewilding degraded landscapes, enforcing protected areas, and implementing wildlife corridors help restore ecosystem balance while reducing human-wildlife conflict and limiting disease transmission risks (Dobson et al., 2020).

Ultimately, ecosystem-based disease management underscores the necessity of preserving and restoring natural habitats as a means of promoting global health security. By safeguarding ecosystems, *Conservation Medicine* not only protects biodiversity but also strengthens public health infrastructure, ensuring a sustainable and disease-resilient future for both humans and wildlife.

1.3.2.3 Biodiversity Conservation as a Health Indicator.

Biodiversity conservation plays a critical role in maintaining ecosystem balance and serves as a fundamental indicator of overall ecosystem health. A diverse and well-functioning ecosystem provides essential services, including air and water purification, climate regulation, and disease suppression. One of the most significant contributions of biodiversity conservation to public and environmental health is its ability to act as a natural buffer against disease transmission, a concept known as the *dilution effect* (Keesing et al., 2010).

The *dilution effect* suggests that in ecosystems with high biodiversity conservation, the presence of a wide range of species reduces the transmission of pathogens by diluting the reservoir of highly competent hosts. In contrast, biodiversity loss often leads to an increased dominance of species that are more efficient at transmitting diseases, thereby amplifying the risk of outbreaks (Ostfeld & Keesing, 2012). For example, studies have shown that in areas with high mammalian diversity, Lyme disease transmission is lower because various host species compete with and dilute the population of white-footed mice, the primary reservoir of *Borrelia burgdorferi*, the bacterium that causes Lyme disease (Levi et al., 2012).

Similarly, deforestation and habitat destruction contribute to biodiversity loss, forcing species into closer proximity with human populations and increasing the likelihood of zoonotic spillover events. Emerging infectious diseases such as Ebola, Nipah virus, and SARS-CoV-2 have been linked to human encroachment into previously undisturbed habitats, where decreased biodiversity conservation has disrupted ecosystem balances and facilitated pathogen transmission (Jones et al., 2008). By protecting and restoring biodiversity, *Conservation Medicine* aims to mitigate these risks and enhance ecosystem resilience.

Beyond disease regulation, biodiversity conservation serves as a health indicator by reflecting the overall morphology, functionality, and species composition of an ecosystem. Healthy ecosystems with robust species diversity are more adaptable to environmental changes and better equipped to recover from disturbances such as climate fluctuations, pollution, and natural disasters. The presence— or decline—of the cornerstone species within an ecosystem, such as pollinators, apex predators, and herbivores, provides valuable insight into the health status of an ecosystem and its capacity to support its services to human and animal populations (Dirzo et al., 2014).

Incorporating biodiversity conservation assessments into conservation and public health strategies is essential for promoting long-term ecosystem and human well-being. *Conservation Medicine* leverages biodiversity monitoring as a tool for early disease detection, ecosystem management, and policy development. By prioritizing habitat conservation, sustainable land-use practices, and biodiversity-friendly agriculture, *Conservation Medicine* strengthens the natural defenses of ecosystems against emerging global health existential threats while ensuring the continued provision of critical ecosystem services.

Ultimately, biodiversity conservation serves as a key diagnostic measure of *Planetary Health*. As ecosystems lose their biodiversity, they become more vulnerable to disease outbreaks, climate instability, and resource depletion. Recognizing biodiversity conservation as an essential component of the essential public health functions and services underscores the need for holistic, integrated, and empirically-derived evidence-based conservation efforts that protect both natural environments and human societies.

1.3.2.4 Interdisciplinary and Multisectoral Collaboration in Conservation Medicine.

Interdisciplinary and multisectoral collaboration is fundamental to *Conservation Medicine*, as tackling the complex challenges of biodiversity loss, emerging diseases, and environmental degradation requires diverse expertise. This field unites professionals from human medicine, wildlife biology, veterinary science, epidemiology, environmental science, public health, and policy to develop holistic, sustainable solutions for global health challenges. By integrating knowledge across these domains, conservation medicine fosters innovative strategies that protect ecosystems while promoting human and animal well-being (Aguirre & Ostfeld, 2013).

One of the primary benefits of interdisciplinary and multisectoral collaboration is the ability to integrate diverse scientific and non-scientific backgrounds, experiences, and perspectives to better understand, adapt and mitigate the interconnected threats facing ecosystems and human populations. For example, medical researchers studying zoonotic diseases work alongside wildlife biologists monitoring animal populations to identify early warning signs of pathogen spillover. Similarly, environmental scientists and ecologists contribute crucial data on habitat loss, pollution, and man-made climate change, helping to shape public policies that protect both biodiversity conservation and public health (Destoumieux-Garzón et al., 2018).

A key example of successful interdisciplinary collaboration is the response to zoonotic disease outbreaks, such as Ebola and SARS-CoV-2 (i.e., COVID-19). During these crises, virologists, epidemiologists, ecologists, and public health officials worked together to trace the origins of the diseases, study transmission pathways, and develop strategies for containment. Conservation biologists provided insight into how habitat destruction and wildlife trade could have contributed to the emergence of disease, reinforcing the need for stronger environmental protections (e.g., laws, regulations, etc.) to prevent future pandemics (Daszak et al., 2020).

Beyond disease management, interdisciplinary and multisectoral collaboration also plays a crucial role in designing and implementing sustainable conservation strategies. *Conservation Medicine* practitioners work with agricultural scientists and policymakers to promote biodiversity-friendly farming practices that reduce habitat destruction while ensuring food security. Urban planners collaborate with ecologists and public health experts to design green infrastructure that enhances biodiversity conservation while improving air and water quality in cities. Indigenous knowledge holders and local communities are also vital partners, as their traditional ecosystem knowledge and land stewardship practices contribute valuable insights into sustainable, empirically-driven evidence-based ecosystem management (Berkes, 2017).

Modern technology further enhances interdisciplinary and multisectoral collaboration in conservation medicine. Advances in remote monitoring, geographic information systems (GIS), and artificial intelligence allow scientists and non-scientists from different fields to analyze environmental and health data in real-time, improving predictive modeling for disease outbreaks and habitat degradation (Zinsstag et al., 2011). Open data-sharing platforms and global partnerships, such as the *One Health High-Level Expert Panel* (OHHLEP), facilitate cross-sectoral cooperation, ensuring that knowledge is effectively translated into action at local, national, and international levels (WHO, FAO, OIE, & UNEP, 2021).

Ultimately, interdisciplinary and multisectoral collaboration is essential for addressing the multifaceted challenges of biodiversity conservation and public health. By fostering strong partnerships between scientists, policymakers, healthcare professionals, and local communities, *Conservation Medicine* ensures that solutions are empirically-derived, evidence-based, culturally relevant, and sustainable. As global environmental and health challenges continue to intensify, the need for integrated, collaborative interdisciplinary and multisectoral approaches has never been more urgent.

1.3.3 Conclusion.

Conservation Medicine plays a pivotal role in addressing the complex and interconnected challenges of biodiversity loss, emerging infectious diseases, and environmental degradation. As human activities continue to alter ecosystems at an unprecedented rate—through deforestation, industrialization, pollution, and climate change—the health of wildlife, domestic animals, and human populations becomes increasingly intertwined. The consequences of these disruptions are far-reaching, affecting global food security, air and water quality, disease dynamics, and ecosystem resilience.

By integrating ecosystems and health systems sciences, *Conservation Medicine* provides a comprehensive, interdisciplinary and multisectoral framework to mitigate these threats. This field emphasizes the importance of biodiversity conservation in maintaining ecosystem stability, recognizing that healthy environments serve as natural buffers against the spread of infectious diseases and other ecosystem imbalances. The principles of *Conservation Medicine* align closely with the *One Health* and *Planetary Health* approaches, reinforcing the need for collaborative efforts across human medicine, veterinary science, environmental policy, public health, and community engagement.

To effectively address current and future challenges, conservation medicine must be embedded in global policy frameworks, public health initiatives, and sustainable development strategies. Strengthening international cooperation, investing in research on zoonotic diseases, and promoting habitat conservation efforts are critical steps in safeguarding both natural ecosystems and human well-being. Additionally, fostering partnerships between scientists, policymakers, healthcare providers, and local communities will ensure that conservation efforts are empirically-driven, evidence-based, and culturally inclusive.

In today's rapidly changing world, *Conservation Medicine* offers a proactive, holistic and, integrative approach to sustaining life on Earth. By prioritizing the health of all living organisms and the ecosystems they inhabit, *Conservation Medicine* serves as a crucial bridge between environmental stewardship and public health practices, ensuring a more resilient and environmentally sustainable future for generations to come.

1.4 References.

1. Aguirre, A. A., Ostfeld, R. S., Tabor, G. M., House, C., & Pearl, M. C. (2002). *Conservation medicine: Ecological health in practice*. Oxford University Press.
2. Aguirre, A. A., & Ostfeld, R. S. (2013). Biodiversity and human health: Science and policy connections. *Frontiers in Ecology and the Environment, 11*(5), 244-251.
3. Bellard, C., Cassey, P., & Blackburn, T. M. (2016). Alien species as a driver of recent extinctions. *Biology Letters, 12*(2), 20150623.
4. Berkes, F., Colding, J., & Folke, C. (2000). Rediscovery of traditional ecological knowledge as adaptive management. *Ecological Applications, 10*(5), 1251-1262.
5. Berkes, F. (2017). *Sacred ecology*. Routledge.
6. Ceballos, G., Ehrlich, P. R., & Dirzo, R. (2017). Biological annihilation via the ongoing sixth mass extinction signaled by vertebrate population losses and declines. *Proceedings of the National Academy of Sciences*, 114(30), E6089-E6096.
7. Chaudhary, A., & Brooks, T. M. (2018). Land use intensity and biodiversity loss. *Environmental Research Letters*, 13(7), 074020.
8. Convention on Biological Diversity (CBD). (2020). Global Biodiversity Outlook 5.
9. Daszak, P., Cunningham, A. A., & Hyatt, A. D. (2004). Emerging infectious diseases of wildlife–threats to biodiversity and human health. *Science, 287*(5452), 443-449.
10. Daszak, P., Olival, K. J., & Li, H. (2020). A strategy to prevent future pandemics similar to COVID-19. *Science*, 369(6502), 379-381.

11. Destoumieux-Garzón, D., Mavingui, P., Boetsch, G., et al. (2018). The One Health concept: 10 years old and a long road ahead. *Frontiers in Veterinary Science*, 5, 14.

12. Díaz, S., Settele, J., Brondízio, E. S., et al. (2019). Pervasive human-driven decline of life on Earth points to the need for transformative change. *Science*, 366(6471), eaax3100.

13. Dirzo, R., Young, H. S., Galetti, M., Ceballos, G., Isaac, N. J., & Collen, B. (2014). Defaunation in the Anthropocene. *Science*, *345*(6195), 401-406.

14. Dobson, A. P., Pimm, S. L., Hannah, L., et al. (2020). Ecology and economics for pandemic prevention. *Science, 369*(6502), 379-381.

15. Food and Agriculture Organization (FAO). (2021). One Health: Food and Agriculture Organization of the United Nations.

16. Food and Agriculture Organization (FAO). (2020). The state of world fisheries and aquaculture 2020.

17. Foley, J. A., et al. (2011). Solutions for a cultivated planet. *Nature, 478*(7369), 337-342.

18. Gebreyes, W. A., Dupouy-Camet, J., Newport, M. J., et al. (2014). The global one health paradigm: Challenges and opportunities for tackling infectious diseases at the human, animal, and environment interface in low-resource settings. *PLoS Neglected Tropical Diseases, 8*(11), e3257.

19. Gibbs, E. P. J. (2014). The evolution of One Health: A decade of progress and challenges for the future. *Veterinary Record, 174*(4), 85-91.

20. Gibson, L., et al. (2011). Primary forests are irreplaceable for sustaining tropical biodiversity. *Nature, 478*(7369), 378-381.

21. Hansen, M. C., et al. (2013). High-resolution global maps of 21st-century forest cover change. *Science, 342*(6160), 850-853.

22. Hebert, P. D., et al. (2003). Biological identifications through DNA barcodes. *Proceedings of the Royal Society B: Biological Sciences, 270*(1512), 313-321.

23. Hoegh-Guldberg, O., et al. (2007). Coral reefs under rapid climate change and ocean acidification. *Science, 318*(5857), 1737-1742.

24. *Intergovernmental Panel on Climate Change* (IPCC). (2022). *Climate Change 2022: Impacts, Adaptation, and Vulnerability.* Intergovernmental Panel on Climate Change Report.

25. Jones, K. E., Patel, N. G., Levy, M. A., et al. (2008). Global trends in emerging infectious diseases. *Nature, 451*(7181), 990-993.

26. Karesh, W. B., Dobson, A., Lloyd-Smith, J. O., et al. (2012). Ecology of zoonoses: Natural and unnatural histories. *The Lancet,* 380(9857), 1936-1945.

27. Keesing, F., Belden, L. K., Daszak, P., et al. (2010). Impacts of biodiversity on the emergence and transmission of infectious diseases. *Nature, 468*(7324), 647-652.

28. Levi, T., Kilpatrick, A. M., Mangel, M., & Wilmers, C. C. (2012). Deer, predators, and the emergence of Lyme disease. *Proceedings of the National Academy of Sciences, 109*(27), 10942-10947.

29. Mackenzie, J. S., & Jeggo, M. (2019). The One Health approach—why is it so important? *Tropical Medicine and Infectious Disease, 4*(2), 88.

30. Murray, K. A., & Daszak, P. (2013). Human ecology in pathogenic landscapes: Two hypotheses on how land use change drives viral emergence. *Current Opinion in Virology, 3*(1), 79-83.

31. O'Neill, J. (2016). Tackling drug-resistant infections globally: Final report and recommendations. *Review on Antimicrobial Resistance.*

32. Osofsky, S. A., Cleaveland, S., Karesh, W. B., et al. (Eds.). (2005). *Conservation and development interventions at the wildlife/livestock interface: Implications for wildlife, livestock, and human health.* IUCN.

33. Ostfeld, R. S., & Keesing, F. (2012). Effects of host diversity on infectious disease. *Annual Review of Ecology, Evolution, and Systematics, 43*, 157-182.

34. Parmesan, C., & Yohe, G. (2003). A globally coherent fingerprint of climate change impacts across natural systems. *Nature, 421*(6918), 37-42.

35. Reid, P. C., Hari, R. E., Beaugrand, G., et al. (2019). Global impacts of climate change on freshwater ecosystems. *Hydrobiologia, 768*(1), 1-28.

36. Rockström, J., et al. (2009). A safe operating space for humanity. *Nature, 461*(7263), 472-475.

37. Rosen, G. E., & Smith, K. F. (2010). Summarizing the evidence on the international trade in illegal wildlife. *EcoHealth, 7*(1), 24-32.

38. Scheffers, B. R., De Meester, L., Bridge, T. C., et al. (2016). The broad footprint of climate change from genes to biomes to people. *Science, 354*(6313), aaf7671.

39. Seddon, N., Chausson, A., Berry, P., Girardin, C. A., Smith, A., & Turner, B. (2020). Understanding the value and limits of nature-based solutions to climate change and other global challenges. *Philosophical Transactions of the Royal Society B, 375*(1794), 20190120.

40. Simberloff, D., et al. (2013). Impacts of biological invasions: what's what and the way forward. *Trends in Ecology & Evolution, 28*(1), 58-66.

41. Taylor, L. H., Latham, S. M., & Woolhouse, M. E. J. (2001). Risk factors for human disease emergence. *Philosophical Transactions of the Royal Society B: Biological Sciences, 356*(1411), 983-989.

42. Travis, D. A., Watson, R. P., & Tauer, A. (2011). The spread of pathogens through trade in wildlife. *Revue Scientifique et Technique, 30*(1), 219-239.

43. Wilcox, C., et al. (2015). Threat of plastic pollution to seabirds is global, pervasive, and increasing. *Proceedings of the National Academy of Sciences, 112*(38), 11899-11904.

44. World Health Organization (WHO). (2021). One Health: Joint Plan of Action.

45. WHO, FAO, OIE, & UNEP. (2021). One Health High-Level Expert Panel (OHHLEP) definition and framework. *World Health Organization.*

46. Wunder, S. (2007). The efficiency of payments for environmental services in tropical conservation. *Conservation Biology, 21*(1), 48-58.

47. Zinsstag, J., Schelling, E., Waltner-Toews, D., & Tanner, M. (2011). From 'one medicine' to 'One Health' and systemic approaches to health and well-being. *Preventive Veterinary Medicine, 101*(3-4), 148-156.

2.0

The Foundational Pillars of Conservation Medicine.

2.1 Introduction.

Conservation Medicine is built upon several foundational pillars that integrate ecosystem, veterinary, and human health sciences to address the complex interactions between species and their environments. These pillars include *One Health, Ecosystem Health, Wildlife Conservation, Emerging Infectious Diseases,* and *Environmental Stewardship* (Aguirre et al., 2002). The *One Health* approach (previously introduced), which emphasizes the inextricable interconnectedness of human, animal, and environmental health, has been central to the field's development (Zinsstag et al., 2011). *Ecosystem Health* highlights the role of biodiversity conservation and environmental integrity in maintaining well-being and resiliency (Rapport et al., 1998). *Wildlife Conservation* is crucial, as habitat destruction and species decline are key drivers of *Emerging Infectious Diseases* (Daszak et al., 2000). Additionally, understanding and mitigating zoonotic disease spillover events, such as those linked to deforestation and climate change, form a major focus of *Conservation Medicine* (Carlson et al., 2021). Lastly, *Environmental Stewardship* promotes sustainable policies and practices that protect both public health and biodiversity conservation (Lebov et al., 2017). Together, these five pillars provide a comprehensive framework for addressing global health challenges through a holistic, integrated, comprehensive, interdisciplinary and multisectoral approach.

2.1.1 The Historical Evolution of Conservation Medicine.

2.1.1.1 Introduction.

Conservation Medicine (previously defined) is an interdisciplinary and multisectoral field that examines the inextricably interconnected relationships between human health, animal health, and ecosystem health. This health and science discipline has evolved over time, integrating insights from veterinary medicine, human medicine, ecology, public health, and environmental sciences. Its roots can be traced back to the early observations of disease transmission between animals and humans, but it gained recognition as a distinct health and science discipline in the late 20th and early 21st centuries.

2.1.1.2 Early Foundations: Zoonotic Disease Recognition.

The concept of diseases shared between animals and humans has been documented for centuries. The earliest recorded zoonotic diseases include rabies, recognized as early as 2000 BCE in Mesopotamia, and the bubonic plague, which devastated Europe in the 14th century (McNeill, 1976). During the 19th century, Louis Pasteur's work on rabies and Robert Koch's postulates (i.e., set of four criteria used to establish whether a specific microorganism is the causative agent of a disease) laid the foundation for understanding pathogen transmission between species (Koch, 1884; Pasteur, 1885).

The 20th century saw an increasing awareness of emerging infectious diseases (EIDs) with zoonotic origins, such as influenza pandemics of 1918 (and later in 21st century with COVID-19 pandemic) and vector-borne diseases (i.e., illnesses transmitted to humans and animals by vectors which are living organisms, such as mosquitos, ticks, and fleas, that carry pathogens like bacteria, viruses, and parasites) like malaria, yellow fever, and West Nile fever. The recognition of the role of wildlife in disease transmission became particularly significant with the emergence of infectious diseases such as Lyme disease (Steere et al., 1977) and the discovery of HIV's origins in primates (Hahn et al., 2000).

2.1.1.3 The Rise of One Health.

In the late 20th century, the *One Health* approach emerged, advocating for a holistic view of health that considers human, animal, and environmental health together. This major paradigm shift was catalyzed by landmark reports on ecosystem changes and their impact on disease ecology, including the 1989 National Research Council report on emerging infections (Institute of Medicine, 1992).

The formalization of *Conservation Medicine* as a distinct field occurred in the 1990s, largely due to the efforts of scientists such as William Karesh, Peter Daszak, and others who studied the intersections between biodiversity conservation, human health, and animal disease ecology (Daszak et al., 2000). The publication of *Conservation Medicine: Ecological Health in Practice* (Aguirre et al., 2002) was a turning point in the international acceptance of *Conservation Medicine* as a separate and distinct health and scientific discipline, providing a comprehensive framework for integrating ecosystem and health sciences.

2.1.1.4 21st Century Developments and Global Implications.

The 21st century has seen increasing recognition of the impact of human activity on disease emergence. Deforestation, man-made climate change, and wildlife trade have been identified as major drivers of new infectious diseases, including SARS (2003), H1N1 influenza (2009), Ebola (2014-2016), and COVID-19 (2020-2023) (Jones et al., 2008; Carlson et al., 2021). *Conservation Medicine* has played a critical role in understanding these outbreaks by emphasizing the importance of biodiversity conservation and habitat protection in preventing disease spillover events.

Interdisciplinary and multisectoral collaborations between conservation biologists, epidemiologists, veterinarians, and public health officials have expanded the field, leading to initiatives such as the *EcoHealth Alliance* and the *Global Virome Project,* which aim to predict and prevent future pandemics by studying wildlife reservoirs (Carroll et al., 2018).

2.1.1.5 Conclusion.

Conservation Medicine has evolved from early observations of zoonotic diseases into a fully integrated scientific and health discipline that acknowledges the deep interconnections between human, animal, and environmental health. Made up of five distinct fields of study, *One Health, Ecosystem Health, Wildlife Conservation, Emerging Infectious Diseases,* and *Environmental Stewardship, Conservation Medicine* recognizes that ecosystem's disturbances—such as habitat destruction, man-made climate change, and biodiversity loss—have direct consequences on the emergence and spread of infectious diseases, food security, and overall public health (Aguirre et al., 2002).

As global environmental changes accelerate, *Conservation Medicine* will play an increasingly critical role in addressing complex health challenges. The rise of zoonotic pandemics, antimicrobial resistance, and ecosystem degradation necessitates a proactive, interdisciplinary approach that merges veterinary health sciences, human health sciences, epidemiology, public health, ecology, and public policy. By promoting sustainable land-use practices, enhancing disease surveillance, and fostering cross-sector collaboration, *Conservation Medicine* provides a roadmap for mitigating health crises while preserving biodiversity conservation.

Looking ahead, advancements in technology, genomics, and data-driven ecosystem's modeling will further enhance our ability to predict and prevent emerging existential health threats. The future of global health and well-being depends on the continued integration of *Conservation Medicine* principles into policy and practice at all levels of socio-ecological organization, ensuring that human progress aligns with environmental and healthcare stewardship and planetary well-being.

2.1.2 An In-depth Review of the Five Foundational Pillars of Conservation Medicine.

Conservation Medicine (previously defined) is an interdisciplinary and multi-sectoral health and science discipline that integrates human, animal, and environmental health to address global challenges such as biodiversity loss, emerging infectious diseases, and environmental degradation. At its core (previously identified), *Conservation Medicine* is built upon five foundational pillars: *One Health, Ecosystem Health, Wildlife Conservation, Emerging Infectious Diseases,* and *Environmental Stewardship*. Each pillar contributes to a holistic and integrated interdisciplinary and multisectoral approach to sustaining life on Earth while mitigating the negative impacts of human activity.

2.1.2.1 One Health.

Previously defined and discussed, the *One Health* approach recognizes the inextricable interconnectedness of human, animal, and environmental health. This concept is fundamental to *Conservation Medicine*, as it acknowledges that diseases can easily transfer between species and that human health is directly affected and surrounded by the health of the environment and wildlife. *One Health* emphasizes collaboration among medical, veterinary, and ecosystem sciences to prevent and control zoonotic diseases such as COVID-19, Ebola, and avian influenza (Zinsstag et al., 2011). Additionally, *One Health* principles are vital in addressing antimicrobial resistance (AMR), which results from overuse of antibiotics in human and veterinary medicine (Robinson et al., 2016). By promoting sustainable interactions among all living organisms, *One Health* fosters resilience against global health threats.

2.1.2.2 Ecosystem Health.

Ecosystem Health pertains to the stability, resilience, and sustainability of ecosystems. A well-functioning ecosystem provides essential services for both humans and animals to strive and thrive such as clean air and water, climate regulation, and disease regulation (Rapport et al., 1998). *Conservation Medicine* integrates *Ecosystem Health* to assess the impacts of human activity on natural systems, such as deforestation, pollution, and climate change. Human-induced environmental changes often lead to biodiversity loss, which in turn disrupts ecosystem balance and heightens disease risks (Patz et al., 2004). Monitoring ecosystem indicators, including species diversity and habitat integrity, is crucial in maintaining healthy interactions between humans, animals, and their shared environments.

2.1.2.3 Wildlife Conservation.

Wildlife conservation is the third critical pillar in conservation medicine, focusing on the protection of species and their habitats. Habitat destruction, poaching, and illegal wildlife trade pose serious threats to biodiversity conservation (Ripple et al., 2016). Many wildlife species serve as reservoirs for zoonotic pathogens, and human encroachment into their habitats increases the risk of disease spillover events (Karesh et al., 2012). *Conservation Medicine* supports strategies such as protected area management, captive breeding programs, and transdisciplinary research to safeguard endangered species. Conservationists work closely with veterinarians and epidemiologists to study the health of wildlife populations and develop interventions that mitigate disease risks while preserving biodiversity conservation.

2.1.2.4 Emerging Infectious Diseases.

Emerging infectious diseases (EIDs) are novel or rapidly increasing infections that often originate at the human-animal-environmental interface. Many EIDs, including SARS-CoV-2 (e.g., COVID-19), Nipah virus, and Hendra virus, have zoonotic origins (Jones et al., 2008). Deforestation, intensive agriculture, and global travel accelerate the emergence and transmission of these diseases. *Conservation Medicine* employs predictive modeling, genomic surveillance, and field studies to understand the drivers of pathogen spillover and prevent future pandemics. Incorporating disease ecology into biodiversity conservation is essential for the early detection of emerging infectious diseases (EIDs) and the development of rapid, real-time adaptive and mitigation strategies. This integration ensures that public health interventions are aligned with ecosystem preservation, reducing the risk of disease spillover while maintaining ecological balance (Daszak et al., 2001).

2.1.2.5 Environmental Stewardship.

Environmental stewardship involves responsible management of limited natural resources to ensure their sustainability for future generations. This pillar emphasizes ethical and sustainable practices in land use, agriculture, industry, and conservation. Man-made climate change, pollution, and habitat fragmentation threaten ecosystem integrity and biodiversity conservation, making *Environmental Stewardship* essential for resilience against environmental crises (Folke et al., 2004). Community engagement, policy development, research, and education play vital roles in promoting *Environmental Stewardship*. By advocating for public policies that reduce GHG emissions, enhance conservation funding, and support sustainable development, *Conservation Medicine* contributes to a balanced coexistence between human society and the natural world.

2.1.2.6 Conclusion.

Conservation Medicine provides an integrative approach and framework that inextricably links human, animal, and environmental health. The five foundational pillars of *Conservation Medicine—One Health, Ecosystem Health, Wildlife Conservation, Emerging Infectious Diseases, and Environmental Stewardship*—serve as guiding principles for addressing complex health and ecosystem challenges. By fostering interdisciplinary and multisectoral collaboration by promoting sustainable policies and practices, *Conservation Medicine* aims to mitigate emerging existential health threats to all living things while preserving the planet's biodiversity conservation and ecosystem services.

2.2 Planetary Health: Definition, Principles, and Practices.

2.2.1 Definition and Principles of Planetary Health.

Planetary Health is an interdisciplinary field that examines the interdependent relationships between human health, environmental systems, and the broader biosphere. It recognizes that the well-being of human populations is directly linked to the health of natural ecosystems, emphasizing the urgent need for sustainable environmental stewardship to address global health challenges (Whitmee et al., 2015).

Core Principles of Planetary Health include:

1. *Human Health Depends on a Stable Biosphere.*
   The degradation of Earth's natural systems—through deforestation, biodiversity loss, pollution, and man-made climate change—has direct and cascading effects on human health. Changes in air quality, water availability, food security, and exposure to infectious diseases are all influenced by environmental disruptions (Myers, 2017).

47

2. *Man-made Climate Change and Health.*
   Continuously rising global temperatures and more frequent extreme weather events contribute to food and water scarcity, malnutrition, heat-related illnesses, vector-borne diseases, and displacement of vulnerable populations. The increasing frequency of climate-related disasters highlights the need for policies that integrate climate resilience and public health strategies (Watts et al., 2021).

3. *Biodiversity Conservation and Disease Emergence.*
   Ecosystem destruction alters the balance between species, increasing the likelihood of zoonotic disease spillover. Human encroachment into wildlife habitats has been linked to outbreaks such as Ebola, SARS, and COVID-19, emphasizing the need for biodiversity conservation efforts to reduce disease emergence risks (Daszak et al., 2020).

4. *Sustainable Food and Water Systems.*
   Agricultural expansion and the rapid growth of the science-industrial complex have led to deforestation, soil depletion, and freshwater shortages, threatening food security and increasing exposure to harmful chemicals. Sustainable agricultural practices and water resource management are vital for maintaining long-term planetary and human health (Springmann et al., 2018).

5. *Interdisciplinary and Policy-Driven Solutions.*
   Addressing *Planetary Health* challenges requires an interdisciplinary and multisectoral approach that includes policymakers, healthcare professionals, environmental scientists, economists, and local communities. Integrating *Planetary Health* into multi-level governance, urban planning, and international health policies can drive systemic changes that promote both ecosystem and human resilience (Horton et al., 2014).

2.2.2 Future Directions in Planetary Health.

The field of *Planetary Health* is evolving with advancements in environmental monitoring, artificial intelligence, and predictive modeling, which enhances the ability to assess ecosystem changes and their health implications. Strong foundational education and capacity-building initiatives also play a crucial role in the future development of the next generation of professionals with the skills needed to implement sustainable solutions. As societies around the globe face increasing environmental and health crises, *Planetary Health* provides a framework for achieving long-term equilibrium between human civilization and the Earth's life-support systems.

 Planetary Health is well-grounded in its research, education, policy, and practice. The following traits and characteristics help to summarize the disciplines value-proposition to the scientific community and the public-at-large:

1. *Interdependence* – Human health is deeply connected to ecosystems, including air, water, food, and biodiversity conservation (Horton et al., 2014).
2. *Sustainability* – Solutions to global health problems must consider ecosystem and biosphere sustainability to prevent long-term environmental damage (Myers, 2017).
3. *Equity and Justice* – Vulnerable and marginalized populations suffer the most from *Planetary Health* threats, necessitating healthy public policies that prioritize upstream environmental and social justice (Whitmee et al., 2015).
4. *Transdisciplinary Collaboration* – *Planetary Health* integrates insights from medicine, ecology, public health, veterinary sciences, economics, sociology, and public policy to address complex health challenges (Myers & Frumkin, 2020).
5. *Preventive Action* – Emphasis on early intervention to mitigate man-made climate change, pollution, and other environmental

risks before they severely impact health and well-being (Haines et al., 2020).

2.2.3 Practices in Planetary Health.

The implementation of *Planetary Health* principles involves:

1. *Climate-Resilient Healthcare Systems* – Designing health infrastructure to withstand climate threats while reducing GHG emissions (Ebi et al., 2018).
2. *Biodiversity Conservation for Health* – Protecting ecosystems to preserve sources of medicine, water quality, and disease regulation (Myers et al., 2013).
3. *Sustainable Agriculture and Food Systems* – Promoting plant-based diets and regenerative farming to mitigate climate impacts and improve nutrition (Springmann et al., 2018).
4. *Urban Planning for Wellness and Well-being* – Green spaces, active transportation, and pollution control to enhance mental and physical health (Gascon et al., 2016).
5. *Global Governance and Policy Integration* – Aligning international policies for man-made climate change, public health practices, and economic development (Costello et al., 2009).

2.2.4 Planetary Health vs. Conservation Medicine.

While *Planetary Health* takes a broad, human-centered approach, *Conservation Medicine* primarily focuses on the intersection of wildlife health, ecosystem integrity, and human well-being. Previously discussed, *Conservation Medicine* is defined as an emerging health and scientific discipline that examines how environmental changes affect human and animal health, with an emphasis on ecosystem health and balance (Aguirre et al., 2002).

Aspect	*Planetary Health*	*Conservation Medicine*
Focus	Human health within environmental systems	Health of ecosystems, wildlife, and humans
Key Drivers	Climate change, pollution, food systems, urbanization	Habitat loss, wildlife diseases, pollution
Scope	Global, policy-driven, systems-based	Regional/ecosystem-based, research-focused
Disciplinary Approach	Transdisciplinary, including medicine, governance, and economics	Interdisciplinary, primarily biomedical and ecological sciences
Application	Public health, sustainability policies, advocacy	Conservation biology, veterinary sciences, epidemiology

2.2.5 Conclusion.

Planetary Health and Conservation Medicine are two complementary disciplines that emphasize the deep inextricable interconnections between ecosystems and human and animal wellness and well-being. However, they differ in scope, focus, and application.

Planetary Health is a broad, policy-driven movement that seeks to address the health impacts of global environmental changes, including climate change, biodiversity loss, pollution, and resource depletion. It is rooted in the understanding that human health is fundamentally dependent on the stability and sustainability of Earth's natural systems. By integrating public health, environmental science, and policy, *Planetary Health* aims to promote sustainable development, climate resilience, and equitable health outcomes on a global scale.

Conservation Medicine, on the other hand, takes a more focused, interdisciplinary approach to studying the dynamic interactions between human, animal, and ecosystem health. This field is particularly concerned with the health of wildlife populations, emerging zoonotic diseases, and the consequences of habitat destruction and biodiversity loss. *Conservation Medicine* seeks to mitigate disease transmission at the human-animal-environment interface and safeguard ecological balance to prevent future health crises.

While *Planetary Health* provides a macro-level framework for sustainable health solutions, *Conservation Medicine* offers critical insights into specific ecosystem-health interactions, particularly those affecting wildlife and infectious disease emergence. By integrating the perspectives and methodologies of both disciplines, researchers, policymakers, and healthcare professionals can develop more comprehensive strategies to address pressing global health challenges. In an era of rapid environmental change, bridging these fields is essential to ensuring the long-term resilience of both human societies and the ecosystems that sustain them.

2.3 Core disciplines in Conservation Medicine.

2.3.1 Introduction.

Conservation Medicine is a multidisciplinary field that integrates scientific information and knowledge with practical approaches from diverse health and science disciplines to address the complex relationships between human, animal, and environmental health. This inextricable interconnected approach is crucial in mitigating existential threats such as zoonotic disease emergence, habitat destruction, biodiversity loss, and environmental pollution. Several core disciplines contribute to *Conservation Medicine*, including *veterinary medicine, ecology, wildlife biology, essential public health functions and services, environmental health and policy, social sciences and Indigenous knowledge, and human medicine.* Each of these fields plays a vital role in shaping the strategies and interventions necessary to promote *Conservation Medicine* (Aguirre et al., 2012; Atlas et al., 2022).

2.3.2 Veterinary Medicine.

Veterinary Medicine is at the core of *Conservation Medicine*. Veterinarians play a critical role in disease prevention, biodiversity conservation, and ecosystem health, addressing both the direct health needs of animals and the indirect consequences of animal diseases on human populations and the environment.

One of the most pressing challenges veterinarians encounter in *Conservation Medicine* is zoonotic disease prevention and control. More than 70% of emerging infectious diseases (EIDs) world-wide originate from animals, particularly wildlife, with notable examples including Ebola, avian influenza, rabies, and coronaviruses such as SARS, MERS, and COVID-19 (Karesh et al., 2012). By conducting disease surveillance, early detection, and response efforts, veterinarians help mitigate the risk of zoonotic spillover, protecting the public's health and minimizing economic disruptions caused by pandemics.

Wildlife veterinarians are at the forefront of conservation initiatives, working to safeguard threatened and endangered species while monitoring the health of wildlife populations and ecosystems. Their responsibilities include:

1. *Disease Surveillance and Epidemiology*: Tracking infectious diseases in wildlife and assessing their potential to spread to humans and domestic animals (Murray et al., 2016).
2. *Health Monitoring and Species Management*: Conducting health assessments, rehabilitating injured animals, and managing breeding programs for conservation.
3. *Biosecurity and Habitat Protection*: Developing strategies to reduce human-wildlife interactions that facilitate disease transmission, such as regulating ecotourism and livestock-wildlife contact.

Veterinarians also play a pivotal role in reintroducing species into the wild, ensuring that released animals are disease-free and ecologically adapted to their native environments. Programs such as captive breeding and vaccination campaigns for endangered species—such as the black-footed ferret in North America or the mountain gorilla in Africa—demonstrate the impact of veterinary interventions on species survival (Gilbert et al., 2020).

Beyond zoonotic diseases, veterinarians are instrumental in addressing broader *One Health* challenges, including:

1. *Antimicrobial Resistance (AMR)*
   a) Overuse and misuse of antibiotics in livestock production, aquaculture, and companion animals contribute to the rise of antimicrobial-resistant pathogens.
   b) Veterinarians lead antibiotic stewardship programs by promoting responsible antibiotic use, alternative treatments, and improved husbandry practices to reduce reliance on antimicrobials (Destoumieux-Garzón et al., 2018).

2. *Illegal Wildlife Trade and Biodiversity Conservation*
   a) The global wildlife trade is a major driver of biodiversity loss and a key source of disease emergence, with illicit markets facilitating the spread of pathogens across regions.
   b) Veterinarians work in wildlife rescue and rehabilitation, forensic investigations and necropsy-driven cause and manner of death determination, and policy advocacy to combat poaching, trafficking, and habitat destruction (Wyatt et al., 2021).

3. *Environmental Pollutants and Toxicology*
   a) Chemical contaminants, heavy metals, pesticides, and plastics pose significant risks to wildlife health, disrupting endocrine systems and bioaccumulating in food chains.
   b) Veterinary toxicologists monitor pollutant exposure in sentinel species, such as birds of prey and marine mammals, to assess ecosystem health and recommend mitigation strategies (Shore & Rattner, 2021).

As human populations continue to expand into wildlife habitats, the role of veterinarians in *Conservation Medicine* will become even more essential. Strengthening cross-disciplinary collaboration between veterinarians, physicians, ecologists, and policymakers under *One Health* and *Planetary Health* frameworks will be crucial in addressing emerging disease threats, habitat destruction, and climate change-driven health challenges (Osofsky et al., 2020).

Investments in wildlife health infrastructure, disease monitoring networks, and sustainable animal health policies will be necessary to prevent future pandemics, conserve biodiversity, and protect global public health. By integrating veterinary expertise with conservation science, the world can better prepare for ecosystem disruptions and public health crises, ensuring a more sustainable future for both humans and wildlife.

2.3.3 Ecology.

Ecology provides the theoretical and empirical foundation for understanding the complex interactions between living organisms and their environment, making it an essential discipline within *Conservation Medicine.* By studying the ways in which man-made climate change, habitat fragmentation, deforestation, and land-use changes impact biodiversity conservation and ecosystem stability, ecologists play a crucial role in identifying emerging health threats and developing strategies for mitigation (Daszak et al., 2000).

One of the most pressing ecological concerns is the human-driven alteration of ecosystems, which has profound consequences for species survival, disease dynamics, and ecosystem services. For example, rapid urban expansion and agricultural intensification have been linked to increased human-wildlife contact, facilitating zoonotic spillover events—a key driver of emerging infectious diseases (EIDs) such as Ebola, Nipah virus, and coronaviruses (Allen et al., 2017). Additionally, climate change-driven shifts in temperature, precipitation, and habitat availability have altered the distribution of vector-borne diseases, such as malaria and Lyme disease, exposing new populations to these threats (Patz et al., 2004).

Ecosystem research is crucial for developing predictive models that assess the risk of disease outbreaks, biodiversity loss, and ecosystem service disruptions. Ecologists apply empirically-derived evidence-based approaches to understand how environmental changes influence disease transmission dynamics, allowing for the early detection and prevention of potential epidemics (Morse et al., 2012).

Some key ecological approaches to disease risk assessment include:

1. *Ecological Niche Modeling*: Predicting how changes in climate and land use will affect the geographic range of disease vectors and reservoirs.
2. *Biodiversity Conservation and Dilution Effect Studies*: Examining whether higher species diversity can reduce pathogen transmission by decreasing the abundance of competent hosts (Ostfeld & Keesing, 2012).
3. *Ecosystem Service Assessments:* Evaluating how the degradation of natural systems impacts water purification, pollination, and carbon sequestration, all of which have implications for human and animal health (Myers et al., 2013).

Restoration Ecology, which focuses on rehabilitating degraded ecosystems, is integral to promoting long-term health outcomes across species. By restoring habitats and improving ecosystem function, conservation scientists can enhance biodiversity conservation, reduce disease transmission, and build resilience to man-made climate change (Chazdon, 2008).

Key restoration strategies include:

1. *Reforestation and Afforestation:* Restoring forests to reduce GHG emissions, stabilize temperatures, and restore wildlife habitats essential for ecological balance.

2. *Wetland Restoration*: Enhancing natural flood control, water purification, and disease vector regulation (e.g., reducing mosquito breeding grounds to limit malaria transmission).

3. *Soil Regeneration and Sustainable Agriculture*: Implementing regenerative farming techniques to maintain soil microbiomes that contribute to crop resilience, food security, and reduced reliance on chemical pesticides.

By integrating ecological principles into *Conservation Medicine* and public health strategies, researchers and policymakers can develop sustainable solutions that address both biodiversity conservation and global health challenges. The application of ecosystem-based approaches will be essential in preventing future pandemics, mitigating climate-related health risks, and ensuring environmental resilience for future generations.

2.3.4 Wildlife Biology.

Wildlife Biology is a critical discipline within *Conservation Medicine*, focusing on the study of animal populations, their behaviors, genetic diversity, and disease ecology. Wildlife biologists play a crucial role in assessing the impact of environmental changes, such as pollution, invasive species, climate change, and habitat destruction, on animal health, biodiversity conservation, and ecosystem stability (Dobson et al., 2003). Their work is essential in understanding and mitigating the threats posed by human activities to wildlife populations and in developing strategies to promote species resilience in changing environments.

A fundamental component of *Wildlife Biology* in *Conservation Medicine* is the study of disease ecology—how pathogens interact with animal hosts and their environments. Wildlife biologists investigate:

1. *Disease transmission dynamics:* Understanding how pathogens spread between wildlife, domestic animals, and humans, partic-

ularly in the context of zoonotic diseases such as avian influenza, rabies, and chronic wasting disease.

2. *Environmental stressors and immune response:* Examining how pollution, climate stress, and habitat loss weaken wildlife immune systems, making populations more susceptible to disease.

3. *Reservoir species identification:* Determining which species harbor and transmit infectious agents, helping to prevent future pandemics through targeted surveillance and intervention (Travis et al., 2011).

Wildlife biologists are deeply involved in species conservation, working to protect endangered species and their natural habitats. Conservation efforts include:

1. *Health monitoring and disease prevention:* Conducting regular health assessments of vulnerable populations to detect emerging health threats.

2. *Captive breeding and reintroduction programs:* Managing breeding initiatives for species such as the California condor, black-footed ferret, and giant panda, ensuring genetic diversity and preparing populations for successful reintegration into the wild (Wobeser, 2006).

3. *Mitigating human-wildlife conflicts:* Developing solutions to reduce conflicts between humans and large carnivores, migratory species, and urban-adapted wildlife through strategies like wildlife corridors, community education, and habitat restoration (Treves & Karanth, 2003).

The rise of the *One Health* approach has led to the increased integration of *Wildlife Biology* into global disease surveillance and environmental health initiatives. Wildlife biologists contribute to early detection and rapid response programs, ensuring that disease outbreaks in wildlife are identified before they spread to livestock and human populations. Examples of *One Health* collaborations include:

1. *Avian Influenza Monitoring:* Tracking waterfowl migration patterns to predict the spread of H5N1 and H7N9 viruses and implementing biosecurity measures to protect poultry industries.
2. *Bat Surveillance Programs:* Studying bat populations to assess their role as reservoirs for coronaviruses, Ebola, and Nipah virus, leading to risk assessments and mitigation strategies.
3. *Climate-Driven Disease Shifts:* Examining how rising temperatures and habitat changes alter the range of disease vectors such as ticks and mosquitoes, influencing malaria, Lyme disease, and West Nile virus prevalence (Patz et al., 2004).

As human activities continue to alter natural ecosystems, the role of *Wildlife Biology* in *Conservation Medicine* will become even more vital. Future research and conservation efforts must focus on:

1. *Advancing Genetic and Molecular Tools* – Using whole genomic sequencing and environmental DNA (eDNA) monitoring to detect wildlife diseases and population trends more effectively.
2. *Enhancing Multisector Collaboration* – Strengthening partnerships between wildlife biologists, veterinarians, epidemiologists, and conservationists to develop holistic strategies for biodiversity conservation and public health protection.
3. *Implementing Adaptive Conservation Strategies* – Designing climate-resilient protected areas and habitat corridors that allow species to adapt to environmental changes while minimizing human-wildlife conflicts.
4. *Expanding Public Engagement and Policy Advocacy* – Promoting science-based conservation policies and increasing public awareness about the ecological consequences of deforestation, illegal wildlife trade, and unsustainable resource use.

By integrating wildlife health research, conservation science, and public health initiatives, *Wildlife Biology* continues to be a cornerstone of *Conservation Medicine*, ensuring that ecosystems remain resilient and that emerging health threats are addressed in a proactive and sustainable manner.

2.3.5 Essential Public Health Functions and Services.

2.3.5.1 Introduction.

Conservation Medicine is an emerging, interdisciplinary field that integrates human, animal, and environmental health under the *One Health* framework. It recognizes the inextricable interconnectedness of ecosystems and the role of public health in preventing zoonotic diseases, mitigating environmental health risks, and ensuring biodiversity conservation (Aguirre et al., 2012). Within this context, the *Essential Public Health Functions (EPHFs)* and *Essential Public Health Services (EPHSs)* serve as a foundation for implementing effective public health interventions that align with conservation efforts.

2.3.5.2 Essential Public Health Functions (EPHFs).

The *EPHFs* are fundamental actions required to improve, promote, and protect public health. The World Health Organization (WHO) and the Pan American Health Organization (PAHO) define these functions as key competencies that any socio-ecological organization level public health system must maintain (PAHO, 2020). Within *Conservation Medicine*, these functions take on additional dimensions due to the interactions between humans, wildlife, and ecosystems.

1. *Monitoring and Assessment of Population Health and Environmental Hazards.*
   a) Surveillance systems must be designed to monitor zoonotic diseases, biodiversity loss, and environmental pollutants affecting health (Jones et al., 2008).

b) Integrated surveillance models, such as the Predict Project, track emerging infectious diseases (EIDs) in wildlife and their transmission potential to humans (Daszak et al., 2000).

2. *Health Protection Through Policy and Regulation.*
   a) Governments and international agencies must establish regulatory frameworks that balance public health protection with conservation goals (WHO, 2021).
   b) Examples include CITES (Convention on International Trade in Endangered Species of Wild Fauna and Flora), which restricts the wildlife trade to prevent zoonotic spillover (Smith et al., 2017).

3. *Disease Prevention and Control.*
   a) Vaccination programs for both humans and animals, such as the oral rabies vaccination of wildlife, help reduce cross-species transmission (Rupprecht et al., 2008).
   b) Control strategies for vector-borne diseases, such as malaria and Lyme disease, require habitat modification and vector population management (Kilpatrick & Randolph, 2012).

4. *Health Promotion and Community Engagement.*
   a) Public health campaigns must incorporate messages on biodiversity conservation, sustainable agriculture, and wildlife interactions to reduce ecosystem and health adverse risks (Mazet et al., 2009).
   b) Indigenous and rural communities play a crucial role in *Conservation Medicine* strategies, necessitating culturally-competent education and training (Karesh et al., 2012).

5. *Emergency Preparedness and Response.*
   a) *Conservation Medicine* requires rapid response mechanisms for outbreaks with environmental origins (e.g., Ebola, H5N1) (Plowright et al., 2017).

b) Disaster preparedness must integrate ecosystem risk assessments and man-made climate change mitigation strategies (Patz et al., 2005).

6. *Stewardship and Governance.*
   a) Strong leadership is needed to coordinate public health agencies, conservation organizations, and veterinary services in a *One Health* approach (Osofsky et al., 2005).
   b) Cross-sectoral policies should guide sustainable development while safeguarding ecosystem services that support human health (Myers et al., 2013).

7. *Research, Innovation, and Evidence-Based Decision Making.*
   a) *Conservation Medicine* relies on transdisciplinary research to evaluate human-wildlife interactions, antimicrobial resistance, and land-use changes impacting disease emergence (Gibb et al., 2020).
   b) Advancements in metagenomics and AI-driven epidemiological modeling improve early disease detection (Carroll et al., 2018).

2.3.5.3 Essential Public Health Services (EPHSs).

The *EPHSs*, defined by the CDC and revised in 2020, represent a framework for delivering public health at local, national, and global levels (CDC, 2020). Within the realm of *Conservation Medicine*, these services ensure sustainable health practices that integrate ecosystem services considerations.

1. *Assess and Monitor Population Health Status and Risks.*
   a) Disease surveillance programs must incorporate wildlife, domestic animal, and human health indicators to detect emerging health threats (Morse et al., 2012).

b) Technologies like environmental DNA (eDNA) help track pathogen reservoirs in aquatic ecosystems (Thomsen & Willerslev, 2015).

2. *Investigate, Diagnose, and Address Environmental and Health Hazards.*
   a) Multi-sectoral teams must analyze deforestation, habitat encroachment, and illegal wildlife trade as drivers of zoonotic spillovers (Daszak et al., 2020).
   b) Field epidemiologists conduct ecosystem niche modeling to predict outbreak hotspots (Peterson, 2014).

3. *Communicate Effectively with Cultural Humility and Empathy to Inform and Educate.*
   a) Conservation health professionals develop public health messaging using cultural humility and empathy when communicating to the local indigenous community on responsible wildlife interactions, antibiotic resistance, and sustainable food systems (Schwabe, 1984).
   b) Mobile health (*mHealth*) platforms provide rapid information on wildlife-borne diseases to rural communities (Bonwitt et al., 2018).

4. *Strengthen, Support, and Mobilize Communities.*
   a) Empowering indigenous groups and local conservationists improves wildlife stewardship and zoonotic risk reduction (Hernandez et al., 2020).
   b) Programs like *EcoHealth Alliance* train communities in biodiversity conservation and disease surveillance and response (Osofsky et al., 2005).

5. *Create and Champion Policies that Support Public and Environmental Health.*
   a) Laws restricting wildlife markets, land conversion, and deforestation help mitigate human-animal disease interfaces (Wolfe et al., 2007).
   b) The *One Health Global Network* coordinates policy efforts across environmental, agricultural, and health sectors (WHO, 2021).

6. *Utilize Legal and Regulatory Actions to Protect Health and Ecosystems.*
   a) Environmental regulations such as the *Clean Water Act* prevent waterborne disease outbreaks linked to industrial pollution (Gibbs, 2014).
   b) Wildlife protection laws, such as *IUCN* Red List classifications, guide conservation efforts to prevent ecosystem disruptions (Baillie et al., 2004).

7. *Assure an Effective Health and Conservation Workforce.*
   a) Training and retaining *One Health* professionals enhances interdisciplinary and multisectoral collaboration in outbreak response and conservation (Rabinowitz et al., 2013).
   b) Veterinary public health programs integrate ecosystem health into clinical practice (Zinsstag et al., 2011).

8. *Improve and Innovate Public and Conservation Health Practices.*
   a) Novel diagnostic tools, including CRISPR-based biosensors, provide rapid pathogen detection in remote settings (van Doremalen et al., 2020).
   b) Sustainable development initiatives, such as agroecology, reduce human encroachment on wildlife habitats (Foley et al., 2011).

9. *Build and Maintain a Strong Public Health Infrastructure.*
   a) International collaborations, such as the *Global Virome Project*, establish a pandemic prediction framework rooted in *Conservation Medicine.* (Carroll et al., 2018).
   b) Strengthening primary health care systems in biodiversity conservation hotspots mitigates ecosystem health risks (Mills et al., 2010).

2.3.5.4 Conclusion.

Essential Public Health Functions and Services are critical in the interdisciplinary and multisectoral approach of *Conservation Medicine.* As human activities increasingly disrupt ecosystems, the integration of public health principles and practices with environmental stewardship becomes indispensable for preventing disease emergence, promoting sustainable development, and protecting biodiversity conservation. A global commitment to *One Health* principles ensures that *Essential Public Health Functions and Services* align with *Conservation Medicine* goals, ultimately fostering a healthier planet for all species.

2.3.6 Environmental Health and Policy.

Environmental Health is a critical field that examines how environmental factors—such as pollution, chemical contaminants, deforestation, industrial waste, and man-made climate change—affect human and ecosystem health, well-being, and long-term resiliency. The degradation of natural ecosystems through air and water pollution, soil contamination, biodiversity loss, and resource depletion directly contribute to the global burden of disease. Exposure to hazardous environmental conditions has been linked to respiratory and cardiovascular diseases, neurological disorders, cancer, and infectious disease transmission, disproportionately affecting vulnerable populations, including children, the elderly, and economically disadvantaged communities (Frumkin, 2016).

Man-made climate change exacerbates these challenges by altering weather patterns, increasing the frequency of natural disasters, and intensifying heatwaves, droughts, and flooding, all of which have profound health consequences. Rising temperatures contribute to heat-related illnesses, worsening air quality, and the expansion of vector-borne diseases such as malaria, dengue, and Lyme disease. Additionally, habitat destruction disrupts ecosystems, driving wildlife closer to human settlements and increasing the risk of zoonotic disease spillover events, as seen with outbreaks of Ebola, SARS, and COVID-19 (Watts et al., 2018).

Environmental Policy plays a pivotal role in safeguarding the public's health by enacting and enforcing regulations that limit hazardous exposures, reduce disease transmission, and promote sustainable resource management. Effective policies address key environmental determinants of health, such as:

1. *Air Quality Regulations:* Reducing emissions from industrial processes, transportation, and energy production to prevent respiratory and cardiovascular diseases.
2. *Water and Sanitation Standards:* Ensuring access to clean drinking water and safe sanitation systems to prevent waterborne illnesses such as cholera and typhoid.
3. *Toxic Chemical Controls:* Regulating the use of pesticides, heavy metals, and persistent organic pollutants that can bioaccumulate in ecosystems and harm human health.
4. *Antibiotic and Pesticide Use Policies:* Controlling excessive agricultural antibiotic use to combat antimicrobial resistance (AMR) and limiting harmful pesticide exposure that affects both human populations and biodiversity (Landrigan et al., 2018).
5. *Sustainable Land Use and Biodiversity Conservation:* Protecting forests, wetlands, and marine environments to maintain ecological balance and reduce the risk of disease emergence.

Governments, international organizations, and scientific communities recognize the urgent need for integrated, empirically-derived evidence-based policies that address environmental health challenges holistically. Institutions such as the World Health Organization (WHO), the United Nations Environment Programme (UNEP), and the Intergovernmental Panel on Climate Change (IPCC) advocate for comprehensive policy frameworks that consider health-related social needs (HRSNs), environmental justice, economic equity, and the commercial determinants of health (Myers et al., 2013). These efforts emphasize a precautionary approach—prioritizing prevention over reactive measures—to build resilience in both human populations and ecosystems.

Multilateral agreements, such as the Paris Agreement on Climate Change, the Minamata Convention on Mercury, and the Stockholm Convention on Persistent Organic Pollutants, demonstrate the importance of international cooperation in reducing environmental hazards. Additionally, *One Health* and *Planetary Health* frameworks emphasize the interconnectivity of environmental, human, and animal health, advocating for cross-sector collaboration in addressing emerging global health threats (Whitmee et al., 2015).

To mitigate the adverse health impacts of environmental degradation, policymakers must implement strategies that:

1. *Strengthen Climate Adaptation and Mitigation Measures –* Investing in clean energy, sustainable agriculture, and resilient infrastructure to reduce climate-related health risks.
2. *Promote Green Technologies and Circular Economies –* Encouraging waste reduction, renewable energy, and sustainable industrial practices to minimize ecological harm.
3. *Enhance Public Health Surveillance and Early Warning Systems –* Developing monitoring frameworks to detect emerging health threats linked to environmental changes.

4. *Prioritize Environmental Justice* – Addressing disparities in exposure to pollution and environmental hazards, particularly in marginalized communities.

5. *Foster Global Collaboration* – Expanding international partnerships to share best practices, research, and resources for sustainable development.

By integrating public health expertise with environmental governance, societies can build resilient, health-conscious public policies that not only mitigate harm but also promote long-term sustainability for future generations. As environmental challenges continue to evolve, a proactive, interdisciplinary and multisectoral approach will be essential in ensuring a healthier planet for all.

2.3.7 Social Sciences and Indigenous Knowledge.

Conservation Medicine benefits greatly from the social sciences, including anthropology, sociology, and economics. These disciplines provide essential insights into the cultural, behavioral, and economic drivers that influence environmental degradation, disease emergence, and public health outcomes. By analyzing human interactions with ecosystems, social scientists help identify the ways in which deforestation, habitat destruction, poaching, intensive agriculture, and urbanization contribute to biodiversity loss and the spread of zoonotic diseases (Berkes, 2018).

For instance, anthropologists' study *Traditional Ecological Knowledge (TEK)*, cultural practices, and human-wildlife interactions, revealing how shifts in land use, migration patterns, and dietary habits affect environmental sustainability and disease risk. Sociologists examine the social structures, policies, and inequities that shape environmental behaviors, such as the role of poverty, urbanization, and industrialization in increasing reliance on unsustainable resource extraction. Meanwhile, economists assess the costs and benefits of conservation efforts, investigating how market forces, trade policies, and incentives influence land use decisions, biodiversity protection, and health disparities (Pretty, 2011).

Indigenous Knowledge (IK) and *Traditional Ecological Knowledge (TEK)* are critical components of *Conservation Medicine*, as many Indigenous communities have developed long-standing, place-based practices that promote biodiversity conservation, ecosystem resilience, and sustainable resource management. Indigenous cultures have historically maintained deep ecological knowledge regarding species interactions, climate patterns, medicinal plants, and land stewardship, offering valuable insights into disease prevention and ecosystem health (Reid et al., 2021).

Indigenous communities often practice sustainable hunting, rotational farming, and forest management techniques that minimize habitat destruction and help control disease reservoirs. For example:

1. *The Kayapo people of the Amazon* utilize traditional agroforestry methods that enhance biodiversity conservation while reducing soil degradation.
2. *The Maasai pastoralists in East Africa* manage livestock grazing in a way that maintains rangeland health and prevents overgrazing, reducing the risk of zoonotic disease transmission.
3. *Australian Aboriginal fire management practices* have long been used to prevent large-scale wildfires and maintain ecosystem balance.

Engaging with *Indigenous Knowledge* systems and social science research enables policymakers, scientists, and healthcare professionals to develop culturally competent, locally relevant, and ethically responsible *Conservation Medicine* strategies. Co-management approaches, where Indigenous leaders and conservation scientists collaborate, have shown success in protected area governance, wildlife conservation, and disease monitoring (Garnett et al., 2018).

However, challenges remain in ensuring that *Indigenous Knowledge* is respected and equitably incorporated into conservation policies. Ethical considerations include:

1. *Recognizing Indigenous sovereignty and rights* over traditional lands and resources.
2. *Ensuring fair compensation and acknowledgment* for Indigenous contributions to conservation efforts.
3. *Avoiding knowledge extraction without community-led decision-making.*
4. *Bridging Indigenous and Western scientific knowledge* through mutual respect and dialogue rather than imposing external conservation models.

By integrating social sciences and Indigenous Knowledge into *One Health* and *Conservation Medicine* frameworks, global health and environmental sustainability efforts can become more inclusive, effective, and culturally grounded. Addressing environmental health challenges requires recognizing the interconnectedness of human societies, ecosystems, and health, making interdisciplinary and multisectoral collaboration essential in tackling emerging threats to biodiversity conservation and public health.

2.3.8 Human Medicine.

Human medicine plays a pivotal role in *Conservation Medicine*. Physicians and medical researchers contribute significantly to understanding and addressing the direct and indirect health effects of environmental changes, including those driven by deforestation, pollution, habitat destruction, and man-made climate change (McMichael, 2013). These environmental stressors not only threaten biodiversity conservation but also increase the risk of human diseases by altering pathogen transmission dynamics, expanding the range of disease vectors, and reducing access to clean water, nutritious food, and air quality (Haines et al., 2009).

Climate change exacerbates global health challenges by intensifying extreme weather events, worsening air pollution, and altering the distribution of infectious diseases. For instance, rising temperatures and changing precipitation patterns have expanded the range of vector-borne diseases such as malaria, dengue fever, and Lyme disease, increasing their incidence in previously unaffected populations (Ogden et al., 2014). Additionally, environmental degradation contributes to respiratory conditions, cardiovascular diseases, and malnutrition, particularly in vulnerable communities that rely on natural ecosystems for sustenance and livelihoods (Watts et al., 2015).

The integration of human medicine with veterinary and ecological sciences strengthens the *One Health* approach, which acknowledges the interconnectedness of human, animal, and environmental health (Destoumieux-Garzón et al., 2018). Physicians, veterinarians, public health experts, and ecologists must work together to monitor and mitigate health risks arising from biodiversity loss, ecosystem disruption, and zoonotic spillover events. This collaboration is particularly crucial in preventing and responding to zoonotic disease outbreaks, such as Ebola, SARS, MERS, and COVID-19, which have underscored the necessity of a unified global health strategy (Daszak et al., 2000).

Furthermore, antimicrobial resistance (AMR) poses a growing threat at the intersection of human and animal health. Overuse and misuse of antibiotics in both human medicine and livestock production contribute to the emergence of resistant pathogens, endangering medical treatments worldwide. *Conservation Medicine* emphasizes responsible antibiotic stewardship, improved surveillance systems, and sustainable agricultural practices to curb AMR and ensure food security for future generations (Karesh et al., 2012).

By fostering interdisciplinary and multisectoral collaboration, human medicine enhances biodiversity conservation efforts by promoting policies that prioritize environmental health as a key determinant of public health. Integrating medical expertise into biodiversity conservation initiatives can drive innovations in disease surveillance, public health preparedness, and sustainable resource management, ultimately contributing to global health resilience in the face of climate change and environmental degradation (Myers et al., 2017).

2.4 Interdisciplinary Integration in Conservation Medicine.

The integration of multiple disciplines and sectors within *Conservation Medicine* is essential for addressing the increasingly complex and complicated health challenges that arise at the intersection of human, animal, plant, and environmental health. By combining expertise from *veterinary medicine, ecology, wildlife biology, public health, environmental policy, social sciences, Indigenous knowledge, and human medicine*, researchers, educators, and practitioners can develop holistic, integrated, and comprehensive interventions that promote health equity and environmental stewardship and sustainability.

The inextricable interconnectedness of bio- and eco-systems demands a transdisciplinary and multisectoral approach that goes beyond traditional silos of knowledge. Veterinarians play a pivotal role in disease surveillance, response, and wildlife health management, while ecologists and wildlife biologists provide crucial insights into habitat loss, biodiversity conservation, and ecosystem function. Public health professionals, at all socio-ecological levels, contribute essential knowledge on disease surveillance and response, risk mitigation, adaptation and communication, advocacy and public policy implementation, whereas environmental policy experts help shape laws and regulations that mitigate the impact of man-made climate change, pollution, and land-use changes.

Moreover, social sciences and Indigenous knowledge provide invaluable perspectives on community-led conservation actions, ethical considerations, and the cultural dimensions of environmental stewardship. Indigenous communities, with their deep-rooted ecosystem knowledge, have successfully managed biodiversity conservation for centuries, and their inclusion in conservation strategies is critical for sustainable outcomes. Human medicine bridges the gap between environmental changes and their direct effects on human health, ensuring that interventions address both immediate and long-term risks posed by environmental degradation.

72

As man-made climate change, deforestation, urbanization, and industrial expansion accelerate environmental transformation, interdisciplinary and multisectoral collaboration will be vital in fostering adaptive and mitigation strategies that enhance ecosystem resilience and promote sustainable global health and well-being. By strengthening cooperation across disciplines, *Conservation Medicine* can offer innovative, real science-driven solutions that not only prevent emerging health threats but also foster a more harmonious relationship between humans and the natural world. The urgency of these challenges underscores the need for integrative, cross-sectoral, and proactive approaches that safeguard the health of all species and the ecosystems upon which they depend (Travis et al., 2011).

2.5 References.

1. Aguirre, A. A., Ostfeld, R. S., Tabor, G. M., House, C., & Pearl, M. C. (2002). *Conservation Medicine: Ecological Health in Practice.* Oxford University Press.
2. Aguirre, A. A., Ostfeld, R. S., Tabor, G. M., House, C., & Pearl, M. C. (2012). *Conservation Medicine: Ecological Health in Practice.* Oxford University Press.
3. Aguirre, A. A., Daszak, P., & Ostfeld, R. S. (2012). New directions in conservation medicine: Applied cases of ecological health. Oxford University Press.
4. Allen, T., Murray, K. A., Zambrana-Torrelio, C., et al. (2017). Global hotspots and correlates of emerging zoonotic diseases. *Nature Communications, 8,* 1124.
5. Atlas, R. M., Maloy, S. R., & Mercer, C. L. (2022). *One Health: People, Animals, and the Environment.* ASM Press.
6. Baillie, J. E. M., Hilton-Taylor, C., & Stuart, S. N. (2004). *2004 IUCN Red List of Threatened Species: A Global Species Assessment.* IUCN.

7. Berkes, F. (2018). *Sacred Ecology: Traditional Ecological Knowledge and Resource Management*. Routledge.

8. Bonwitt, J., Dawson, M., Kandeh, M., et al. (2018). Unintended consequences of the 'bushmeat ban' in West Africa during the 2013-2016 Ebola virus disease epidemic. *Social Science & Medicine, 200*, 166-173.

9. Carlson, C. J., Albery, G. F., Merow, C., et al. (2021). Climate change increases cross-species viral transmission risk. *Nature*, 607, 555–562.

10. Carroll, D., Daszak, P., Wolfe, N. D., et al. (2018). The Global Virome Project. *Science*, 359(6378), 872-874.

11. Centers for Disease Control and Prevention (CDC). (2020). The 10 Essential Public Health Services.

12. Chazdon, R. L. (2008). Beyond deforestation: Restoring forests and ecosystem services on degraded lands. *Science, 320*(5882), 1458-1460.

13. Costello, A., Abbas, M., Allen, A., et al. (2009). Managing the health effects of climate change. *The Lancet, 373*(9676), 1693-1733.

14. Daszak, P., Cunningham, A. A., & Hyatt, A. D. (2000). Emerging infectious diseases of wildlife: Threats to biodiversity and human health. *Science, 287*(5452), 443-449.

15. Daszak, P., Olival, K. J., & Li, H. (2020). A strategy to prevent future pandemics similar to COVID-19. *Science, 369*(6502), 379-381.

16. Destoumieux-Garzón, D., Mavingui, P., Boetsch, G., et al. (2018). The One Health concept: 10 years old and a long road ahead. *Frontiers in Veterinary Science, 5*, 14.

17. Dobson, A., Foufopoulos, J., & Hudson, P. (2003). Pathogens and the conservation of biodiversity. In L. F. Frenkel & D. E. Sherman (Eds.), *The New Encyclopedia of Mammals* (pp. 121–127). Oxford University Press.

18. Ebi, K. L., Anderson, V., Berry, P., et al. (2018). Health risks of climate change: Act now or pay later. *The Lancet, 392*(10163), 2030-2031.

19. Folke, C., Hahn, T., Olsson, P., & Norberg, J. (2004). Adaptive governance of social-ecological systems. *Annual Review of Environment and Resources, 30*(1), 441-473.

20. Foley, J. A., Ramankutty, N., Brauman, K. A., et al. (2011). Solutions for a cultivated planet. *Nature, 478*(7369), 337-342.

21. Frumkin, H. (2016). *Environmental Health: From Global to Local.* John Wiley & Sons.

22. Garnett, S. T., Burgess, N. D., Fa, J. E., et al. (2018). A spatial overview of the global importance of Indigenous lands for conservation. *Nature Sustainability, 1*(7), 369-374.

23. Gascon, M., Triguero-Mas, M., Martínez, D., et al. (2016). Residential green spaces and mortality: A systematic review. *Environmental International, 86*, 60-67.

24. Gibb, R., Redding, D. W., Chin, K. Q., Donnelly, C. A., Blackburn, T. M., Newbold, T., & Jones, K. E. (2020). Zoonotic host diversity increases in human-dominated ecosystems. *Nature, 584*(7821), 398-402.

25. Gibbs, H. K. (2014). Environmental regulation and public health: The Clean Water Act's impact. *Annual Review of Public Health, 35*, 187-204.

26. Gilbert, M., Lubroth, J., & Veterinary Public Health. (2020). Wildlife health and conservation: Challenges and future directions. *Revue Scientifique et Technique, 39*(1), 61-74.

27. Hahn, B. H., Shaw, G. M., De Cock, K. M., & Sharp, P. M. (2000). AIDS as a zoonosis: scientific and public health implications. *Science*, 287(5453), 607-614.

28. Haines, A., Kovats, R. S., Campbell-Lendrum, D., & Corvalán, C. (2009). Climate change and human health: Impacts, vulnerability, and mitigation. *The Lancet, 367*(9528), 2101-2109.

29. Haines, A., Ebi, K. L., & Smith, K. R. (2020). Climate change and health: Impacts and adaptation. *BMJ, 368*, m810.

30. Hernandez, S. M., Galindo, J., Dyer, T., Burroughs, J. E., & Keel, K. (2020). Integrating indigenous knowledge and conservation medicine for ecosystem health. *EcoHealth, 17*(1), 78-88.

31. Horton, R., & Lo, S. (2014). Planetary health: A new science for exceptional action. *The Lancet, 383*(9928), 2183-2184.

32. Institute of Medicine (IOM) (1992). *Emerging Infections: Microbial Threats to Health in the United States.* National Academies Press.

33. Jones, K. E., Patel, N. G., Levy, M. A., et al. (2008). Global trends in emerging infectious diseases. *Nature,* 451(7181), 990-993.

34. Karesh, W. B., Dobson, A., Lloyd-Smith, J. O., et al. (2012). Ecology of zoonoses: natural and unnatural histories. *The Lancet, 380*(9857), 1936-1945.

35. Kilpatrick, A. M., & Randolph, S. E. (2012). Drivers, dynamics, and control of emerging vector-borne zoonotic diseases. *The Lancet, 380*(9857), 1946-1955.

36. Koch, R. (1884). Die Aetiologie der Tuberkulose. *Berliner Klinische Wochenschrift,* 21, 221-230.

37. Landrigan, P. J., Fuller, R., Acosta, N. J. R., et al. (2018). The Lancet Commission on pollution and health. *The Lancet, 391*(10119), 462-512.

38. Lebov, J., Grieger, K., Womack, D., et al. (2017). A framework for One Health research. *Health Security,* 15(1), 53-63.

39. Mazet, J. A. K., Clifford, D. L., Coppolillo, P. B., et al. (2009). A "One Health" approach to address emerging zoonoses: The HALI Project in Tanzania. *PLoS Medicine, 6*(12), e1000190.

40. McMichael, A. J. (2013). Globalization, climate change, and human health. *New England Journal of Medicine, 368*(14), 1335-1343.

41. McNeill, W. H. (1976). *Plagues and Peoples.* Anchor Press.

42. Mills, J. N., Gage, K. L., & Khan, A. S. (2010). Potential influence of climate change on vector-borne and zoonotic diseases: A review and proposed research plan. *Environmental Health Perspectives, 118*(11), 1507-1514.

43. Morse, S. S., Mazet, J. A., Woolhouse, M., et al. (2012). Prediction and prevention of the next pandemic zoonosis. *The Lancet, 380*(9857), 1956-1965.

44. Murray, M. H., Becker, D. J., Hall, R. J., & Hernandez, S. M. (2016). Wildlife health and supplemental feeding: A review and management recommendations. *Biological Conservation, 204*, 163-174.

45. Murray, K. A., Olivero, J., Roche, B., et al. (2016). Pathogeography: Leveraging the biogeography of human infectious diseases for global health management. *Ecological Health, 13*(4), 490-494.

46. Myers, S. S., Gaffikin, L., Golden, C. D., et al. (2013). Human health impacts of ecosystem alteration. *PNAS, 110*(47), 18753-18760.

47. Myers, S. S. (2017). Planetary health: Protecting nature to protect ourselves. Island Press.

48. Myers, S. S., Gaffikin, L., Golden, C. D., et al. (2017). Human health impacts of ecosystem alteration. *Proceedings of the National Academy of Sciences, 110*(47), 18753-18760.

49. Myers, S. S., & Frumkin, H. (2020). Planetary health: Protecting nature to protect ourselves. *The New England Journal of Medicine, 382*(8), 748-754.

50. Ogden, N. H., Radojevic, M., Wu, X., et al. (2014). Estimated effects of projected climate change on the basic reproductive number of the Lyme disease vector *Ixodes scapularis*. *Environmental Health Perspectives, 122*(6), 631-638.

51. Osofsky, S. A., Cleaveland, S., Karesh, W. B., et al. (2005). *Conservation and Development Interventions at the Wildlife/ Livestock Interface: Implications for Wildlife, Livestock, and Human Health*. IUCN.

52. Osofsky, S. A., Kock, R. A., & Kock, M. D. (2020). One Health and conservation medicine. *Encyclopedia of Biodiversity, 2*, 183-195.

53. Ostfeld, R. S., & Keesing, F. (2012). Effects of host diversity on infectious disease. *Annual Review of Ecology, Evolution, and Systematics, 43*(1), 157-182.

54. Pan American Health Organization (PAHO). (2020). The Essential Public Health Functions in the Americas: A Renewal for the 21st Century.

55. Pasteur, L. (1885). Méthode pour prévenir la rage après morsure. *Comptes Rendus de l'Académie des Sciences*, 101, 765-772.

56. Patz, J. A., Daszak, P., Tabor, G. M., et al. (2005). Unhealthy landscapes: Policy recommendations on land use change and infectious disease emergence. *Environmental Health Perspectives*, *112*(10), 1092-1098.

57. Peterson, A. T. (2014). *Mapping disease transmission risk: Enriching models using biogeography and ecology.* Johns Hopkins University Press.

58. Plowright, R. K., Becker, D. J., McCallum, H., & Manlove, K. R. (2017). Sampling to elucidate the dynamics of infections in reservoir hosts. *Philosophical Transactions of the Royal Society B: Biological Sciences, 372*(1722), 20160036.

59. Pretty, J. (2011). Interdisciplinary progress in approaches to address social-ecological and ecocultural systems. *Environmental Conservation, 38*(2), 127-139.

60. Rabinowitz, P. M., Conti, L., & Yee, J. (2013). *One Health and the Politics of Antimicrobial Resistance.* Johns Hopkins University Press.

61. Rapport, D. J., Costanza, R., & McMichael, A. J. (1998). Assessing ecosystem health. *Trends in Ecology & Evolution*, 13(10), 397-402.

62. Rattner, B. A. (2009). History of wildlife toxicology. *Ecotoxicology, 18*(1), 1-5.

63. Reid, A. J., Eckert, L. E., Lane, J. F., et al. (2021). "Two-Eyed Seeing": An Indigenous framework to transform fisheries research and management. *Fish and Fisheries, 22*(2), 243-261.

64. Ripple, W. J., Newsome, T. M., Wolf, C., et al. (2016). Collapse of the world's largest herbivores. *Science Advances, 2*(5), e1500936.

65. Robinson, T. P., Bu, D. P., Carrique-Mas, J., Fèvre, E. M., Gilbert, M., Grace, D., ... & Van Boeckel, T. P. (2016). Antibiotic resis-

tance is the quintessential One Health issue. *Transactions of the Royal Society of Tropical Medicine and Hygiene, 110*(7), 377-380.

66. Rupprecht, C. E., Hanlon, C. A., & Slate, D. (2008). Oral vaccination of wildlife against rabies: Opportunities and challenges in prevention and control. *Developmental Biology, 131*, 173-184.

67. Schwabe, C. W. (1984). *Veterinary Medicine and Human Health*. Williams & Wilkins.

68. Shore, R. F., & Rattner, B. A. (2021). *Ecotoxicology of wild mammals*. John Wiley & Sons.

69. Smith, K. M., Machalaba, C., Seifman, R., et al. (2017). Infectious disease and economics: The case for considering multi-sectoral impacts. *One Health, 3*, 34-40.

70. Springmann, M., Clark, M., Mason-D'Croz, D., et al. (2018). Options for keeping the food system within environmental limits. *Nature, 562*(7728), 519-525.

71. Steere, A. C., Malawista, S. E., Hardin, J. A., et al. (1977). Erythema chronicum migrans and Lyme arthritis. *Annals of Internal Medicine*, 86(6), 685-698.

72. Thomsen, P. F., & Willerslev, E. (2015). Environmental DNA–An emerging tool in conservation for monitoring past and present biodiversity. *Biological Conservation, 183*, 4-18.

73. Travis, D. A., Watson, R. P., & Tauer, A. (2011). The spread of pathogens through trade in wildlife. *Revue Scientifique et Technique de l'OIE, 30*(1), 219-239.

74. Treves, A., & Karanth, K. U. (2003). Human-carnivore conflict and perspectives on carnivore management worldwide. *Conservation Biology, 17*(6), 1491-1499.

75. van Doremalen, N., Bushmaker, T., Morris, D. H., et al. (2020). Aerosol and surface stability of SARS-CoV-2 as compared with SARS-CoV-1. *The New England Journal of Medicine, 382*(16), 1564-1567.

76. Watts, N., Adger, W. N., Agnolucci, P., et al. (2015). Health and climate change: Policy responses to protect public health. *The Lancet, 386*(10006), 1861-1914.

77. Watts, N., Amann, M., Arnell, N., et al. (2018). The 2018 report of the Lancet Countdown on health and climate change: Shaping the health of nations for centuries to come. *The Lancet, 392*(10163), 2479-2514.

78. Watts, N., Amann, M., Arnell, N., et al. (2021). The 2021 report of The Lancet Countdown on health and climate change: Code red for a healthy future. *The Lancet, 398*(10311), 1619-1662.

79. Whitmee, S., Haines, A., Beyrer, C., et al. (2015). Safeguarding human health in the Anthropocene epoch: Report of the Rockefeller Foundation-Lancet Commission on planetary health. *The Lancet, 386*(10007), 1973-2028.

80. Wilcox, B. A., & Ellis, B. (2006). Forests and emerging infectious diseases of humans. *Unasylva, 57*(224), 11-18.

81. Wobeser, G. (2006). *Essential Veterinary Epidemiology of Wildlife.* Blackwell Publishing.

82. Wobeser, G. (2006). Essential veterinary epidemiology for wildlife biologists. *Journal of Wildlife Diseases, 42*(3), 511-518.

83. Wolfe, N. D., Dunavan, C. P., & Diamond, J. (2007). Origins of major human infectious diseases. *Nature, 447*(7142), 279-283.

84. World Health Organization (WHO). (2021). One Health High-Level Expert Panel to advise on One Health issues.

85. Wyatt, T., van Uhm, D., & Nurse, A. (2021). The illegal wildlife trade: Scale, processes, and governance. *Annual Review of Criminology, 4*, 159-179.

86. Zinsstag, J., Schelling, E., Waltner-Toews, D., & Tanner, M. (2011). From "one medicine" to "One Health" and systemic approaches to health and well-being. *Preventive Veterinary Medicine*, 101(3-4), 148-156.

3.0

Threats to Global Biodiversity Conservation and Planetary Health.

3.1 Introduction.

Biodiversity conservation—the management and biosecurity of the variety of life on Earth, including species, genetic, and ecosystem diversity—is fundamental to *Planetary Health*. It provides essential ecosystem services such as air and water purification, climate regulation, food security, and disease control (Díaz et al., 2019). However, global biodiversity conservation is facing an unprecedented crisis, with species extinction rates estimated to be 100 to 1,000 times higher than natural background rates due to human activities (Ceballos et al., 2015). This rapid biodiversity loss is not only an ecosystem services issue but also a significant threat to global health and well-being for all living things, as it disrupts the delicate balance of interconnected and harmonious ecosystems, increases the risk of emerging infectious diseases (EIDs), and exacerbates food and water insecurity.

Several key drivers are responsible for the decline in biodiversity conservation and the subsequent threats to *Planetary Health*. These include man-made climate change, habitat destruction, pollution, overexploitation of natural resources, and the spread of invasive species (IPBES, 2019). Man-made climate change alters temperature and precipitation patterns, affecting species distributions and ecosystem stability, while deforestation and land-use changes fragment habitats, leading to biodiversity loss and increased human-wildlife interactions that facilitate zoonotic disease spillover (Myers et al., 2013; Daszak et al., 2000). Pollution, including chemical contaminants, plastics, and agricultural runoff, degrades water and soil quality, harming both biodiversity conservation and human health (Landrigan et al., 2018). Overfishing, poaching, and unsustainable agricultural practices further deplete species populations, disrupting food chains and ecosystem services. Additionally, the introduction of non-native species can outcompete or prey on native species, leading to ecosystem imbalances that threaten biodiversity conservation (Simberloff et al., 2013).

The loss of biodiversity conservation has far-reaching consequences for human and animal health. Declining biodiversity conservation reduces the availability of medicinal resources, undermines food security by affecting pollination and soil fertility, and increases the spread of infectious diseases through ecological disruption (Keesing et al., 2010). Emerging infectious diseases, including SARS, Ebola, and COVID-19, have been linked to biodiversity loss and habitat encroachment, highlighting the critical relationship between ecosystem integrity and global health (Jones et al., 2008).

Addressing these threats requires an interdisciplinary and multisectoral approach to develop sustainable solutions. *Conservation Medicine* and the *One Health* framework emphasize the inextricable interconnectedness of human, animal, and ecosystem health, advocating for public policies that protect biodiversity conservation while promoting global health security (Destoumieux-Garzón et al., 2018). As environmental changes continue to accelerate, an urgent call to action is needed to mitigate biodiversity loss and its cascading effects on *Planetary Health*.

3.2 Habitat Destruction and Deforestation.

3.2.1 Introduction.

Habitat destruction and deforestation are among the most significant drivers of biodiversity loss, posing existential threats to global health, ecological stability, and human well-being. As forests are cleared for agriculture, urban expansion, and industrial development, countless species lose their natural habitats, leading to population declines and, in many cases, extinction. This loss of biodiversity disrupts critical ecosystem services such as pollination, water purification, and climate regulation, ultimately destabilizing ecosystems that support life on Earth.

Moreover, deforestation and habitat fragmentation create conditions that facilitate the emergence and transmission of zoonotic diseases—infectious diseases that jump from animals to humans. As wildlife is forced into closer contact with human populations due to habitat loss, the risk of pathogen spillover increases, as seen in outbreaks of diseases such as Ebola, Lyme disease, and COVID-19. Additionally, reduced biodiversity conservation can lead to an imbalance in predator-prey dynamics, allowing disease-carrying species, such as rodents and mosquitoes, to thrive unchecked.

Recognizing and addressing the intricate connections between deforestation, biodiversity loss, and human health is essential for developing sustainable conservation strategies. Protecting and restoring natural habitats, promoting sustainable land-use practices, and integrating public health with environmental policies can mitigate these risks. A global commitment to reforestation, habitat preservation, and responsible resource management is critical to safeguarding biodiversity conservation and ensuring a healthier, more resilient future for both ecosystems and human populations.

3.2.2 Biodiversity Loss Due to Habitat Destruction.

Biodiversity refers to the vast array of life on Earth, encompassing species diversity, genetic variation, and the complexity of ecosystems. This biological richness supports the stability and functionality of the natural world, playing a crucial role in sustaining ecosystem services that support human well-being. However, biodiversity is increasingly threatened by human activities, with deforestation ranking among the most significant drivers of ecosystem degradation and species loss.

Deforestation, largely fueled by agricultural expansion, logging, and infrastructure development, leads to the fragmentation of habitats and the displacement of wildlife (Newbold et al., 2016). As forests are cleared, many species lose their natural homes, disrupting ecosystem interactions and leading to population declines. Additionally, the destruction of forests diminishes the resilience of ecosystems, making them less capable of recovering from environmental disturbances. This degradation compromises essential ecosystem services such as carbon sequestration, water filtration, and soil stabilization, which are vital for maintaining climate balance and supporting agricultural productivity (Foley et al., 2005).

Tropical forests, which contain approximately 80% of the planet's terrestrial species, are disproportionately affected by deforestation, making them a focal point for conservation efforts (Gibson et al., 2011). The Amazon rainforest, the world's largest tropical forest, has experienced widespread deforestation due to cattle ranching, soy cultivation, and illegal logging. This has placed immense pressure on iconic species such as jaguars (Panthera onca), harpy eagles (Harpia harpyja), and giant river otters (Pteronura brasiliensis), all of which depend on intact forest ecosystems' services for survival. Habitat fragmentation further exacerbates biodiversity loss by isolating populations, reducing genetic diversity, and increasing species' vulnerability to environmental changes (Haddad et al., 2015).

Addressing deforestation requires a multisectoral approach that includes strengthening conservation policies, promoting sustainable land-use practices, and fostering international collaboration. Reforestation initiatives, protected area expansion, and responsible supply chain management can help mitigate biodiversity loss and restore ecosystem function. By recognizing the interconnectedness of biodiversity conservation, climate stability, and human well-being, policymakers and conservationists can work collaboratively toward preserving the Earth's rich biological heritage for future generations.

3.2.3 The Impact of Habitat Destruction on Human Health.

Habitat destruction has profound and complex consequences on human health, both directly and indirectly. One of the most critical concerns is the emergence and spread of zoonotic diseases—infectious diseases that originate in animals and have the potential to spill over into human populations. As deforestation accelerates, wildlife is forced into closer proximity to human settlements, increasing opportunities for novel pathogens to cross species barriers (Jones et al., 2008). This ecosystem disruption has been implicated in several major disease outbreaks, including the Ebola virus, which has been linked to forest fragmentation and human encroachment in Central Africa. Similarly, COVID-19 is suspected to have originated from the wildlife trade and habitat degradation, highlighting the intricate interconnections between ecosystem health and global public health (Gibb et al., 2020).

Beyond zoonotic disease transmission, deforestation plays a significant role in exacerbating man-made climate change by releasing vast amounts of stored carbon into the atmosphere. Forests act as crucial carbon sinks, absorbing and storing carbon dioxide (CO_2) that would otherwise contribute to rising global temperatures. When forests are cleared, this carbon is released, intensifying the greenhouse effect and leading to rising global temperatures (Malhi et al., 2008). These climatic shifts drive more frequent and severe weather events, such as hurricanes, droughts, and wildfires, which have devastating effects on human health and infrastructure.

Additionally, man-made climate change driven by deforestation alters disease dynamics, particularly for vector-borne diseases like malaria and dengue fever. Warmer temperatures and changing precipitation patterns create favorable conditions for disease-carrying mosquitoes, expanding their geographic range and increasing transmission rates in previously unaffected regions (Patz et al., 2005). This shift places millions of people at greater risk, straining essential public health functions and services and disproportionately affecting vulnerable and marginalized populations in low- and middle-income countries and tropical regions.

Addressing the health risks associated with habitat destruction requires a holistic, interdisciplinary approach that integrates environmental health and biodiversity conservation with essential public health functions and services. Strengthening global efforts to curb deforestation, promoting sustainable land-use policies, and enhancing disease surveillance in high-risk areas are critical steps toward mitigating these existential threats. Recognizing the inextricable interconnectedness of ecosystem integrity, climate stability, and human health and well-being is essential for developing long-term solutions that protect both biodiversity conservation and the public's health and well-being.

3.2.4 Comprehensive Strategies for Addressing Deforestation and Habitat Destruction.

Effectively combating deforestation and habitat destruction requires an interdisciplinary, multisectoral approach that integrates biodiversity conservation, sustainable land-use practices, and evidence-based public policy interventions. Given the complex interplay between economic development, environmental stewardship, and global health, addressing these challenges necessitates coordinated action at local, national, and international levels.

One of the most effective means of mitigating deforestation is the expansion and enforcement of protected areas, which serve as refuges for biodiversity conservation and help maintain essential ecosystem services. Establishing ecosystem pathways between fragmented habitats can further promote species survival and genetic diversity (Chazdon, 2008). Reforestation and afforestation efforts, including large-scale tree-planting initiatives and natural forest regeneration, play a crucial role in restoring degraded landscapes and enhancing carbon sequestration. Additionally, agroforestry—a practice that integrates trees and vegetation into agricultural systems—can help balance food production with environmental stewardship, reducing the need for forest clearance while maintaining soil health and water retention.

Indigenous and local communities have long been effective stewards of biodiversity-rich ecosystems, managing forests sustainably for generations. Research shows that deforestation rates are significantly lower in regions where land rights are legally recognized and indigenous governance is upheld (Walker et al., 2020). Empowering these communities through legal protections, financial incentives, and participatory land management programs enhances conservation outcomes while supporting cultural heritage and traditional ecosystem knowledge.

Global environmental policy frameworks play a pivotal role in addressing habitat destruction. International agreements such as the *Convention on Biological Diversity (CBD)* and the *United Nations' Sustainable Development Goals (SDGs)* provide strategic frameworks for countries to develop biodiversity conservation policies, promote sustainable land use, and curb deforestation. National policies, such as strict land-use zoning regulations, aligned financial incentives for biodiversity conservation, and penalties for illegal logging, further reinforce efforts to protect ecosystems' services. Additionally, market-based mechanisms such as *REDD+ (Reducing Emissions from Deforestation and Forest Degradation)* offer financial incentives to developing LMICs to reduce deforestation and enhance sustainable forest management.

Deforestation is often driven by global supply chains, particularly for commodities such as palm oil, soy, beef, and timber.

Corporate commitments to sustainable sourcing, along with consumer awareness and advocacy, can significantly reduce pressure on forests. Certifications such as the *Forest Stewardship Council (FSC)* and *Roundtable on Sustainable Palm Oil (RSPO)* help ensure responsible production practices, while consumer choices—such as reducing palm oil consumption and supporting ethically sourced forestry products—contribute to conservation efforts (Lambin et al., 2018).

Advancements in satellite monitoring, remote sensing, and artificial intelligence have improved the ability to track deforestation in real time, enabling governments and conservation organizations to respond swiftly to illegal activities. Scientific research on forest ecology, carbon sequestration, and ecosystem resilience continues to inform empirically-derived evidence-based best practices for reforestation and sustainable land management. Additionally, biotechnology innovations, such as developing drought-resistant crops and alternative proteins, can reduce the need for land conversion while ensuring food security.

Mitigating deforestation and habitat destruction requires a holistic, collaborative approach that aligns environmental stewardship and biodiversity conservation with economic and social priorities. By integrating ecosystem restoration, indigenous stewardship, policy reform, corporate accountability, and technological innovation, societies can create sustainable pathways that protect biodiversity conservation, combat climate change, and support human health and well-being. The success of these strategies hinges on global multisectoral cooperation and long-term commitment to preserving the planet's natural ecosystems for future generations.

3.2.5 Conclusion.

Habitat destruction and deforestation are among the most pressing threats to global biodiversity conservation and human health, driven primarily by agricultural expansion, urbanization, logging, and infrastructure development. These processes not only fragment ecosystems and accelerate species extinction but also disrupt ecosystem services critical for maintaining environmental homeostasis. The loss of forests, wetlands, and other natural habitats reduces carbon sequestration, exacerbating man-made climate change, while simultaneously diminishing biodiversity conservation and threatening the survival of countless species.

Beyond ecosystem consequences, deforestation and habitat destruction pose significant public health challenges world-wide. As human populations encroach on previously undisturbed habitats (e.g., urbanization), the likelihood of zoonotic disease spillover increases, facilitating the emergence and spread of infectious diseases such as Ebola, Lyme disease, and COVID-19. The substantial reduction of natural barriers between humans and wildlife enhances contact with novel pathogens, heightening the risk of global pandemics. Furthermore, the degradation of ecosystems' services compromises essential resources, including clean air, water, and medicinal plants, which are vital for human well-being.

Addressing these urgent threats requires the implementation of sustainable biodiversity conservation strategies that balance economic development with environmental stewardship. Expanding protected areas, enforcing anti-deforestation policies, and promoting reforestation initiatives are essential steps in preserving biodiversity conservation. Agroforestry, sustainable land-use planning, and responsible supply chain management can mitigate the potentially-adverse environmental impact of agriculture and industry while supporting local economies.

International collaboration is crucial for scaling-up biodiversity conservation efforts and ensuring long-term ecosystem resilience. Governments, scientists, businesses, and communities must work together to develop environmentally-friendly public policies that promote habitat restoration, reduce carbon emissions, and protect endangered species. Integrating traditional ecological knowledge with modern biodiversity conservation practices can also enhance sustainable management of Earth's deteriorating and diminishing natural resources.

By prioritizing environmental biodiversity conservation and sustainable development, societies can mitigate the devastating effects of habitat destruction and deforestation. Protecting biodiversity conservation is not only a moral and ecological imperative but also a necessary strategy for safeguarding the public's health and ensuring a resilient, thriving planet for future generations.

3.3 Man-made Climate Change: A Major Driver of Ecosystem Disruption and Public Health Crises.

3.3.1 Introduction.

Anthropogenic (man-made) climate change is one of the most profound and far-reaching environmental challenges of the modern era, driving widespread disruptions to both ecosystems and public health. The rapid increase in GHG emissions—primarily from fossil fuel combustion, deforestation, and industrial activities—has led to rising global temperatures, altered precipitation patterns, and an increase in the frequency and severity of extreme weather events. These environmental changes are fundamentally reshaping ecosystems, forcing species to migrate in search of suitable habitats and altering the balance of natural eco-systems (IPCC, 2021).

As temperatures rise and climate conditions shift, many species are experiencing habitat loss and fragmentation, reducing their ability to adapt and survive. Marine ecosystems, for example, are suffering from coral bleaching due to ocean acidification and warming waters, leading to declines in biodiversity conservation and fisheries productivity (Hoegh-Guldberg et al., 2017). Similarly, terrestrial species, including pollinators such as bees and butterflies, are being displaced from their native ranges, threatening food security and agricultural stability (Pecl et al., 2017). These disruptions not only endanger biodiversity conservation but also undermine ecosystem services such as carbon sequestration, water purification, and soil fertility, which are essential for human survival.

One of the most pressing public health concerns linked to man-made climate change is its role in the emergence and spread of infectious diseases. As ecosystems shift and wildlife migrates into new areas, the likelihood of zoonotic disease spillover—where pathogens jump from animals to humans—increases. This dynamic has been observed in several disease outbreaks, including the spread of Lyme disease, which has expanded due to warmer temperatures allowing tick populations to thrive in new regions (Ogden et al., 2014). Similarly, vector-borne diseases such as malaria, dengue fever, and Zika virus are spreading to higher altitudes and latitudes, placing previously unaffected populations at risk (Ryan et al., 2019).

Extreme weather events, including hurricanes, floods, and droughts, further compound public health risks by displacing communities, contaminating water supplies, and creating conditions favorable for outbreaks of waterborne diseases such as cholera and leptospirosis. Heatwaves, another consequence of man-made climate change, have been linked to increased mortality rates, particularly among vulnerable and marginalized populations such as the elderly, children, and those with preexisting health conditions (Vicedo-Cabrera et al., 2021).

Given the inextricable interconnected nature of man-made climate change, biodiversity loss, and public health, addressing these challenges requires an integrated, multidisciplinary approach. Strengthening climate adaptation and mitigation strategies—including reducing GHG emissions, enhancing ecosystem resilience, and improving global disease surveillance—is crucial to minimizing the risks associated with climate change-driven health threats.

International agreements such as the *Paris Agreement* and the *United Nations' Sustainable Development Goals (SDGs)* emphasize the urgent need for coordinated global action to mitigate rising temperatures and protect vulnerable communities. Policy interventions at national and local levels—such as investing in renewable energy, promoting climate-smart agriculture, and improving urban resilience to extreme weather—can help mitigate climate impacts while fostering sustainable development.

Additionally, advancing research on climate-sensitive diseases, expanding early-warning surveillance systems, and strengthening healthcare service delivery infrastructure in high-risk regions are essential steps in safeguarding the public's health. Public awareness and community engagement also play a critical role in driving behavioral changes that contribute to climate resilience, such as reducing carbon footprints, supporting biodiversity conservation initiatives, and advocating for climate policies.

As the man-made climate crisis continues to accelerate, recognizing its far-reaching effects on ecosystems and human health is imperative. A proactive, empirically-driven and evidence-based approach to climate adaptation and mitigation can help prevent further biodiversity loss, reduce disease burdens, and protect the health and well-being of future generations.

3.3.2 Species Migration and Biodiversity Loss.

As climate conditions shift, many species are compelled to migrate to maintain access to optimal temperature and resource conditions. This phenomenon is particularly evident in terrestrial and marine ecosystems. Studies show that terrestrial species are shifting their ranges poleward at an average rate of 17 km per decade and upward in elevation by 11 meters per decade (Chen et al., 2011). In marine environments, species migration rates are even more pronounced, with an average shift of 72 km per decade (Poloczanska et al., 2013).

These shifts disrupt existing ecosystems' services, leading to competition, predation imbalances, and potential extinctions. For example, as Arctic ice melts, polar bears (*Ursus maritimus*) are forced southward, increasing conflicts with human populations. Similarly, warming oceans have led to coral reef bleaching, affecting countless marine species dependent on these habitats for life on Earth (Hughes et al., 2017).

3.3.3 Species Migration and Infectious Disease Spread.

One of the most concerning effects of species migration is the increased spread of infectious diseases. Warmer temperatures and altered precipitation patterns create favorable conditions for disease-carrying organisms, particularly mosquitoes, ticks, and rodents, to expand their geographic range.

1. *Vector-borne diseases*: Rising temperatures have allowed malaria-transmitting *Anopheles* mosquitoes to expand into higher altitudes and previously temperate regions (Ryan et al., 2019). Similarly, the incidence of dengue fever has increased in regions that were once too cold for *Aedes aegypti* mosquitoes (Messina et al., 2019).
2. *Zoonotic diseases*: Habitat changes and biodiversity loss are increasing human-wildlife contact, facilitating the emergence of zoonotic diseases such as Ebola, Nipah virus, and coronaviruses (Gibb et al., 2020). As species migrate, pathogens move with them, raising the likelihood of cross-species transmission.
3. *Tick-borne diseases*: The northward expansion of tick populations due to warmer winters has led to a rise in Lyme disease cases in North America and Europe (Ogden et al., 2014).

3.3.4 The Impact of Species and Pathogen Migration on Global Health.

The migration of species, including disease-carrying vectors and pathogens, poses a significant and growing threat to global health. As man-made climate change alters environmental conditions, many organisms are forced to shift their geographic ranges in search of suitable habitats. This redistribution disrupts existing ecosystem balances, alters infectious disease dynamics, and places additional strain on healthcare delivery systems worldwide. Vector-borne diseases, in particular, are becoming endemic in new regions, requiring local and regional public health authorities to rapidly adapt to emerging existential health threats and implement effective disease surveillance and control measures (Myers et al., 2017).

Rising global temperatures, changing precipitation patterns, and habitat modifications are enabling the expansion of disease-carrying organisms such as mosquitoes, ticks, and rodents. For instance, Aedes mosquitoes—primary vectors of dengue fever, Zika virus, and chikungunya—are now establishing populations in higher latitudes and altitudes, increasing the risk of outbreaks in regions previously unaffected (Ryan et al., 2019). Similarly, Lyme disease, transmitted by black-legged ticks, has expanded into northern regions of North America and Europe due to milder winters and extended tick activity seasons (Ogden et al., 2014).

In response, local and national public health systems must adapt by strengthening disease surveillance and response, expanding vector control programs, and enhancing healthcare infrastructure in newly affected regions. Early-warning surveillance systems that integrate climate and epidemiological data can help predict outbreaks, allowing for more proactive responses and equitable resource allocation.

Man-made climate change also exacerbates food and water insecurity, increasing malnutrition rates and making marginalized populations more vulnerable to infectious diseases. Prolonged droughts, desertification, and extreme heatwaves reduce crop yields and compromise food production, leading to increased rates of undernutrition, particularly in low-income countries. Malnourished individuals, especially children, are more susceptible to infections such as pneumonia, diarrheal diseases, and tuberculosis, as weakened immune systems make it harder to fight off pathogens (Myers et al., 2017).

Water scarcity and contamination further compound these challenges. Rising temperatures accelerate the proliferation of waterborne pathogens, increasing the risk of diseases such as cholera, dysentery, and giardiasis. Harmful algal blooms, fueled by warming waters and nutrient pollution, can produce toxins that contaminate drinking water supplies and cause severe gastrointestinal and neurological illnesses.

Extreme weather events—including hurricanes, floods, and cyclones—act as catalysts for infectious disease outbreaks by displacing human populations, disrupting sanitation infrastructure, and contaminating water sources. The 2010 cholera outbreak in Haiti, for example, was linked to post-earthquake flooding, which facilitated the proliferation and transmission of *Vibrio cholerae*, the bacterium responsible for the disease (Piarroux et al., 2011). More recently, devastating floods in Pakistan (2022) led to outbreaks of malaria, dengue, and waterborne illnesses, overwhelming healthcare systems and deepening public health crises.

In disaster-prone regions, investing in resilient infrastructure, improving emergency response surveillance systems, and integrating climate adaptation strategies into public health planning are crucial for minimizing the health impacts of extreme weather. Enhanced access to clean water, sanitation, and hygiene (WASH) services can significantly reduce disease transmission following climate-related disasters.

Mitigating the health risks associated with species and pathogen migration requires a *One Health* approach, which recognizes the inextricable interconnectedness of human, animal, and environmental health. Key strategies include:

1. *Strengthening global disease surveillance* to track the spread of vector-borne and zoonotic diseases.
2. *Investing in climate-resilient healthcare delivery systems* to prepare for emerging infectious diseases.
3. *Enhancing food security programs* to reduce malnutrition and improve community resilience.
4. *Expanding vaccination efforts* in regions at risk of climate-induced disease outbreaks.
5. *Reducing GHG emissions* to slow the pace of climate change and its associated health impacts.

As man-made climate change continues to drive species migration and disease emergence, integrating sustainable environmental and public health policies will be essential in preventing future pandemics and ensuring global health security.

3.3.5 Man-made Climate Change: Mitigation and Adaptation Strategies.

Addressing the biodiversity conservation and health impacts of man-made climate change requires a comprehensive approach that integrates biodiversity conservation, public health practices, and environmental and public health policy interventions. Critical strategies include:

1. *Protecting and restoring ecosystems' services*: Preserving biodiversity through conservation efforts can enhance ecosystem resilience and mitigate disease emergence (Dobson et al., 2020).
2. *Climate-informed healthcare planning:* Enhancing healthcare delivery system infrastructure under the Primary Health Care (PHC) framework—through the integration of Advanced Primary Care (APC) services, Essential Public Health Functions (EPHFs), and community-based services—along with local workforce development and training, strengthens emergency preparedness, resilience, and the capacity to respond to emerging infectious diseases, which is vital for adaptation (McMichael, 2015).
3. *Vector control and disease surveillance*: Monitoring and controlling disease vectors through early warning systems, vaccines, and improved sanitation can help prevent outbreaks (Rocklöv & Dubrow, 2020).
4. *International cooperation and policy action*: Global initiatives such as the *Paris Agreement* and the *One Health* approach emphasize the interconnectedness of climate, biodiversity conservation, and health (IPBES, 2020).

3.3.6 Conclusion.

Man-made climate change-induced species migration and the spread of infectious diseases represent profound threats to global biodiversity conservation and human health and well-being. As rising global temperatures, altered precipitation patterns, and shifting habitats force species to move beyond their historical ranges, ecosystems are experiencing significant disruptions. These disruptions can lead to cascading effects, including increased competition for limited resources, displacement of native species, and heightened extinction risks. Simultaneously, as vectors, pathogens, and hosts expand into new areas, the likelihood of novel and re-emerging infectious diseases affecting both wildlife and human populations grows, compounding global health challenges.

Addressing these interconnected crises requires a comprehensive and urgent response that integrates environmental stewardship, biodiversity conservation, public health initiatives, and socio-economic resilience strategies. Biodiversity conservation efforts must prioritize habitat protection, ecosystem restoration, and biodiversity conservation monitoring to support species adaptation and mitigate disruptions to fragile ecosystems' services. Strengthening emergency-response disease surveillance systems and investing in predictive modeling will be crucial in identifying and controlling emerging and existential health threats before they escalate into global crises. Additionally, fostering climate-resilient healthcare delivery system infrastructure, enhancing vector control measures, and promoting sustainable urban planning will help mitigate the health impacts of shifting disease patterns.

Multisectoral collaboration among governments, scientists, healthcare providers, and local communities is essential in developing effective adaptation and mitigation strategies against man-made climate change. Engaging and empowering indigenous and local knowledge, fostering community-based biodiversity conservation, and promoting environmental public policies that reduce carbon emissions will further contribute to long-term solutions. By integrating conservation science with public health strategies and equitable access to healthcare essential services and supply-chain resources, societies can build resilience against the growing threats posed by climate change-induced migration and infectious disease spread. Ensuring a sustainable and healthy future requires decisive global action, where environmental stewardship and human well-being are addressed as inextricably interconnected priorities.

3.4 Emerging Infectious Diseases.

3.4.1 Introduction.

Emerging infectious diseases (EIDs) present a growing and complex threat to global health, biodiversity conservation, and the stability of ecosystems. These diseases, often driven by novel or previously unrecognized pathogens, have surged in frequency due to a combination of human-driven and environmental factors. Man-made climate change, deforestation, rapid urbanization, and the expansion of global trade have created conditions that facilitate the spillover of infectious agents from wildlife to human populations (Daszak et al., 2000).

The intricate and inextricable connections between human, animal, and environmental health underscore the urgent need for a *One Health* approach—an integrated, interdisciplinary, and multisectoral strategy that recognizes how the health of people, animals, and ecosystems is deeply interdependent. Without coordinated global efforts to address these risks through emergency response disease surveillance, early detection, and sustainable environmental policies, the threat of future pandemics will only continue to escalate (Jones et al., 2008).

3.4.2 Emerging infectious diseases (EIDs) Impact on Biodiversity Conservation.

Emerging infectious diseases (EIDs) have become a leading driver of species decline and extinction, posing a significant threat to global biodiversity conservation and ecosystem services stability. These diseases, often introduced or exacerbated by human activity, can cause devastating population losses across a wide range of taxa, leading to cascading ecosystem consequences.

One of the most well-documented examples is the chytrid fungus (*Batrachochytrium dendrobatidis*), which has decimated amphibian populations worldwide. This highly virulent pathogen has been implicated in the decline or extinction of over 200 amphibian species, making it one of the most destructive wildlife diseases ever recorded (Scheele et al., 2019). Similarly, *Pseudogymnoascus destructans*, the fungal pathogen responsible for white-nose syndrome, has devastated bat populations in North America, killing millions. The loss of these key insectivorous species has disrupted ecosystems' services by reducing natural pest control and altering plant pollination dynamics (Frick et al., 2016).

The rapid spread of these and other wildlife diseases is often facilitated by habitat destruction, man-made climate change, and human-mediated movement of pathogens. Deforestation, urban expansion, and land-use changes increase stress on wildlife populations, weaken immune defenses, and create conditions conducive to pathogen spillover. Additionally, rising global temperatures and shifting precipitation patterns can expand the geographic range of infectious agents, further exacerbating species vulnerability (Daszak et al., 2000).

Mitigating the impact of EIDs on biodiversity conservation requires a proactive, interdisciplinary, and multisectoral approach that integrates conservation strategies with disease surveillance and ecosystem resilience-building efforts. Strengthening the *One Health* framework—which emphasizes the inextricable interconnectedness of human, animal, and environmental health—will be critical in addressing the root causes of wildlife disease emergence and preventing future biodiversity losses.

3.4.3 Emerging infectious diseases (EIDs) Threats to Human Health.

Zoonotic emerging infectious diseases (EIDs), which originate in animals and spill over to human populations, pose a profound and persistent existential public health threat on a global scale. These diseases account for the majority of newly emerging pathogens and have demonstrated their capacity to cause widespread morbidity, mortality, and socioeconomic disruption. The COVID-19 pandemic, caused by SARS-CoV-2 virus, provided a stark example of how quickly novel zoonotic viruses can spread worldwide, overwhelming healthcare delivery systems, straining medical supply chains, and inflicting severe economic consequences. With exceptionally high morbidity and mortality across all age groups, COVID-19 underscored the vulnerabilities of even the most advanced public health infrastructures (Wu et al., 2020).

Beyond COVID-19, other zoonotic threats—including Ebola virus, Nipah virus, and highly pathogenic avian influenza strains—demonstrate the unpredictable nature of emerging pathogens and their potential to spark global pandemics. The 2014–2016 Ebola outbreak in West Africa exposed critical gaps in outbreak emergency preparedness, resilience, and response, while Nipah virus, which has a high fatality rate and no specific treatment, continues to pose a serious risk of regional outbreaks with high pandemic potential (Woolhouse & Gowtage-Sequeria, 2005). Meanwhile, avian influenza viruses such as H5N1 and H7N9 have repeatedly crossed the species barrier, raising concerns about future mutations that could enable sustained human-to-human transmission (Gao et al., 2013).

The increasing frequency of zoonotic spillover events is driven by factors such as habitat destruction, wildlife trade, agricultural intensification, and man-made climate change, all of which bring humans, livestock, and wildlife into closer and more frequent contact. Addressing this growing threat requires a proactive, multidisciplinary *One Health* approach that integrates human, animal, and environmental health strategies. Enhancing global disease surveillance, investing in early warning systems, and bolstering rapid response capabilities are critical to reducing the risk and impact of future pandemics.

3.4.4 Drivers of Emerging infectious diseases (EIDs).

Several factors contribute to the rise of EIDs:

1. *Habitat Destruction and Deforestation*: Deforestation increases human-wildlife interactions, facilitating pathogen spillover. For example, the Nipah virus outbreaks in Malaysia were linked to deforestation, which forced fruit bats into closer proximity with pig farms, creating conditions for viral transmission (Daszak et al., 2001).
2. *Man-made Climate Change*: Rising global temperatures and shifting weather patterns world-wide alter the distribution of vectors such as mosquitoes, leading to the expansion of highly-pathogenic EIDs like malaria, dengue, and Zika virus into new regions (Ryan et al., 2019).
3. *Globalization and Trade*: Increased movement of people and goods around-the world has accelerated the spread of pathogens. The rapid transmission of SARS, COVID-19, and monkeypox across continents demonstrates how modern transportation networks facilitate disease emergence (Bogich et al., 2012).

3.4.5 The *One Health* Approach to Emerging Infectious Diseases (EIDs).

Given the intricate and dynamic interactions between human, animal, and environmental health, a *One Health* approach is essential for effectively addressing emerging infectious diseases (EIDs). This holistic and comprehensive framework recognizes that the health of humans, animals, and ecosystems is inextricably linked, requiring coordinated efforts across multiple disciplines to mitigate disease risks.

As previously discussed, this strategy prioritizes disease surveillance, early detection, and rapid response through interdisciplinary collaboration among human healthcare professionals, veterinarians, epidemiologists, and ecosystem scientists (Karesh et al., 2012). By fostering communication and data sharing across these fields, the *One Health* model enhances the ability to identify and contain EID threats before they escalate into full-scale outbreaks.

In addition to improving surveillance, proactive measures such as strengthening biosecurity protocols, minimizing habitat destruction, and enhancing disease monitoring in wildlife populations are critical to preventing future pandemics. Deforestation and land-use changes increase human-wildlife interactions, facilitating pathogen spillover events, while weak biosecurity in livestock and food production systems can accelerate disease transmission (Daszak et al., 2020). Investing in sustainable land management, responsible agricultural practices, and global pathogen tracking networks will be crucial in reducing the risk of future zoonotic disease emergence.

Ultimately, a robust *One Health* approach requires global collaboration and cooperation, environmental and public policy integration, and sustained financial and workforce investment in research and public health infrastructure. By addressing the root causes of EIDs and fostering resilience in humans, wildlife, and ecosystems, we can better safeguard at-risk populations from the next pandemic threat.

3.4.6 Conclusion.

Emerging infectious diseases (EIDs) represent a profound and growing global challenge, posing substantial existential threats to both biodiversity conservation and human health. Their increasing emergence is driven by complex and interrelated factors, including environmental degradation, man-made climate change, urbanization, global travel, determinants of health, and an ever-widening gap in health disparities. EIDs not only disrupt ecosystems by driving species decline and altering biodiversity conservation but also place immense pressure on healthcare delivery systems, economies, and public health infrastructure world-wide.

Addressing the rising threat of EIDs requires a comprehensive, multidisciplinary, and multisectoral approach that integrates environmental, epidemiological, and socioeconomic strategies. Effective disease surveillance, early detection, and rapid response mechanisms are essential to prevent outbreaks from escalating into global crises. Additionally, mitigating habitat destruction, reducing human-wildlife interactions, and addressing climate-driven shifts in disease vectors are critical components of a proactive public health strategy.

Strengthening international collaboration is vital for enhancing pandemic emergency preparedness, improving data-sharing and interoperability of information systems, and fostering coordinated global responses. Investing in *One Health* initiatives—an approach that recognizes the inextricable interconnectedness of human, animal, and environmental health—will be crucial in developing sustainable disease prevention and protection strategies. By prioritizing research, environmental and public policy integration, and multisectoral partnerships, the global scientific and healthcare communities can work toward mitigating the impact of EIDs and ensuring long-term health security for both human and animal populations and ecosystems.

3.5 Pollution and Toxicological Threats.

3.5.1 Introduction.

Pollution and toxicological threats are among the most significant global challenges posing severe risks to biodiversity conservation and human health. The release of harmful substances into air, water, and soil has led to widespread environmental degradation, species decline, and an increase in communicable infectious diseases in humans.

Anthropogenic (i.e., man-made) activities such as industrial GHG emissions, agricultural runoff, and plastic waste accumulation has intensified the impact of these toxic exposures, necessitating urgent intervention through policy reform, sustainable practices, and international cooperation (Landrigan et al., 2018).

3.5.2 Environmental Pollution as a Major Driver of Biodiversity Loss.

Environmental pollution is a critical factor accelerating biodiversity loss worldwide. The accumulation of toxic substances—including heavy metals, pesticides, and industrial chemicals—disrupts the delicate balance of ecosystems, impairing the reproductive, physiological, and behavioral functions of wildlife. These pollutants infiltrate air, water, and soil, creating cascading effects that threaten species survival and ecosystem resilience.

One of the most concerning pollutants is persistent organic pollutants (POPs), such as polychlorinated biphenyls (PCBs). These long-lived chemicals bioaccumulate in the tissues of marine organisms, reaching hazardous concentrations in top predators. Studies have linked PCBs to immune suppression, hormonal imbalances, and reproductive failures in marine mammals, including seals, dolphins, and whales (Desforges et al., 2018). Such disruptions threaten population stability and, in some cases, push species toward extinction.

Similarly, neonicotinoid pesticides, widely used in agriculture, contribute significantly to pollinator decline. Bees, essential for pollination and food security, suffer from impaired navigation, reduced foraging efficiency, and weakened colony health when exposed to these neurotoxic chemicals (Wood & Goulson, 2017). As pollinator populations dwindle, the negative effects ripple across ecosystems, threatening plant diversity and agricultural productivity.

Aquatic ecosystems are particularly vulnerable to pollution-induced toxicological threats. Eutrophication, driven by excessive nutrient runoff from fertilizers and agricultural waste, triggers harmful algal blooms. These blooms deplete oxygen levels, release potent toxins, and disrupt aquatic food webs, leading to mass fish mortality and ecosystem collapse (Smith & Schindler, 2009). Freshwater and marine habitats alike are affected, with consequences that extend to human communities reliant on fisheries and clean water.

Adding to this crisis, plastic pollution has become a pervasive environmental threat. Microplastics, now found in marine organisms across all trophic levels, compromise digestive and reproductive functions and introduce toxic additives into food chains (Rochman et al., 2016). The ingestion of plastic debris by fish, seabirds, and marine mammals not only causes direct harm but also facilitates the bioaccumulation of hazardous substances, further endangering biodiversity and human health.

Addressing the biodiversity conservation crisis necessitates urgent global action to reduce pollution at its source, enforce stricter environmental regulations, and promote sustainable practices. Without significant intervention, pollution will continue to erode ecosystem integrity, threatening both wildlife and the essential services nature provides to humanity.

3.5.3 The Human Health Impact of Environmental Pollutants.

Human exposure to environmental pollutants is a significant public health concern, contributing to a broad spectrum of acute and chronic illnesses. The widespread contamination of air, water, and food by industrial and agricultural activities has led to an increase in respiratory diseases, neurological impairments, and endocrine disruptions, among other serious health effects. These pollutants not only compromise individual well-being but also place immense burdens on healthcare systems worldwide.

Air pollution remains one of the most pervasive threats to human health, primarily driven by industrial emissions, vehicular exhaust, and the combustion of fossil fuels. Fine particulate matter (PM2.5) is particularly dangerous due to its ability to penetrate deep into the lungs and enter the bloodstream. Chronic exposure to PM2.5 has been linked to an increased risk of lung cancer, stroke, and cardiovascular diseases, as well as neurodevelopmental disorders in children (Lelieveld et al., 2020). Airborne pollutants such as nitrogen oxides (NOx) and sulfur dioxide (SO$_2$) exacerbate conditions like asthma and chronic obstructive pulmonary disease (COPD), disproportionately affecting vulnerable populations, including children, the elderly, and individuals with preexisting conditions.

Chemical pollutants, especially heavy metals like lead, mercury, and cadmium, accumulate in human tissues over time, leading to severe health consequences. Lead exposure, historically associated with contaminated water supplies and aging infrastructure, causes irreversible neurological damage, particularly in children, impairing cognitive function and increasing the risk of developmental disorders. Mercury, primarily ingested through contaminated seafood, disrupts the nervous system, leading to cognitive impairments, motor dysfunction, and fetal developmental abnormalities (Grandjean & Landrigan, 2014). Cadmium, found in cigarette smoke and industrial waste, is associated with kidney dysfunction, bone demineralization, and an elevated risk of cancer.

Endocrine-disrupting chemicals (EDCs), including bisphenol A (BPA) and phthalates, interfere with hormone regulation, leading to widespread health issues. These synthetic chemicals, commonly found in plastics, food packaging, and personal care products, mimic or block natural hormones, contributing to reproductive disorders, metabolic dysfunction, and increased cancer risks (Gore et al., 2015). Studies suggest that prenatal exposure to EDCs is linked to early puberty, obesity, and neurodevelopmental disorders, raising concerns about the long-term implications of widespread chemical exposure.

Unsafe drinking water, contaminated with pathogens, heavy metals, and industrial chemicals, remains a major global health challenge. Polluted water sources are responsible for millions of deaths annually, particularly in low-income regions where access to clean water and sanitation is limited. Arsenic contamination in groundwater has been linked to chronic poisoning, skin lesions, cardiovascular disease, and cancer, while fluoride exposure in excess amounts leads to skeletal fluorosis, a painful condition that weakens bones and joints (Ravenscroft et al., 2009). Inadequate water treatment infrastructure also facilitates the spread of waterborne diseases such as cholera, dysentery, and typhoid, exacerbating health disparities.

The pervasive impact of environmental pollutants on human health underscores the urgent need for stronger regulatory policies, pollution mitigation strategies, and public health interventions. Reducing emissions, phasing out hazardous chemicals, and ensuring access to clean water and air are critical steps toward safeguarding the public's health. Without decisive action, the burden of pollution-related diseases will continue to rise, threatening both current and future generations.

3.5.4 Strategies for Addressing Pollution and Toxicological Threats.

Mitigating the impact of pollution on ecosystems' services and human health requires a complex, coordinated effort across local, national, and global levels. Given the complexity of pollution-related challenges, an integrated approach is necessary to regulate industrial emissions, phase out hazardous chemicals, and transition toward sustainable environmental practices.

Governments play a critical role in reducing environmental contamination through robust policies and regulatory frameworks. Key strategies include:

1. *Stricter emissions controls on industrial pollutants*, including sulfur dioxide (SO_2), nitrogen oxides (NOx), and particulate matter (PM2.5), to reduce air pollution and associated health risks.
2. *Banning or restricting hazardous chemicals*, such as persistent organic pollutants (POPs), heavy metals, and endocrine-disrupting chemicals (EDCs), to minimize bioaccumulation in ecosystems.

3. *Promoting sustainable agricultural practices*, including the reduction of synthetic pesticide and fertilizer use, adoption of integrated pest management (IPM), and expansion of organic farming initiatives.

4. *Enforcing clean water regulations*, such as limiting industrial waste discharge into rivers and lakes, ensuring safe drinking water, and remediating contaminated sites.

Successful implementation of these policies requires international cooperation, as pollutants often cross-national boundaries via air and water currents. Global agreements such as the *Paris Agreement*, the *Stockholm Convention on Persistent Organic Pollutants*, and the *Minamata Convention on Mercury* provide frameworks for collective action against pollution.

Poor waste management exacerbates pollution, particularly in urban areas and developing regions. Advancing sustainable waste management strategies is essential to minimizing pollution's impact on ecosystems and human health. Key approaches include:

1. *Plastic reduction initiatives*, such as bans on single-use plastics, extended producer responsibility (EPR) programs, and the development of biodegradable alternatives (Borrelle et al., 2020).

2. *Enhanced recycling infrastructure* to improve waste sorting, collection, and processing, reducing landfill overflow and environmental contamination.

3. *Waste-to-energy technologies*, such as anaerobic digestion and thermal treatment, to convert organic and non-recyclable waste into renewable energy sources.

4. *Circular economy models*, which prioritize reuse, refurbishment, and material recovery, reducing reliance on raw materials and minimizing industrial pollution.

A *One Health* approach—which recognizes the interconnectedness of environmental, human, and animal health—is crucial for tackling toxicological challenges. By integrating cross-disciplinary expertise in medicine, environmental science, and public health, *One Health* strategies can:

1. *Strengthen environmental monitoring systems* to track pollution sources, assess ecological impacts, and develop early warning systems for emerging toxic threats.
2. *Invest in green technologies*, including renewable energy, eco-friendly industrial processes, and bioremediation techniques to mitigate pollution at its source.
3. *Enhance public awareness and education* through campaigns that inform communities about pollution risks, encourage sustainable consumption patterns, and promote behavioral changes.

Addressing pollution and toxicological threats demands sustained global action, innovative policy frameworks, and investments in sustainable technologies. Without decisive intervention, pollution will continue to erode biodiversity, threaten public health, and exacerbate climate change. A collaborative, science-driven approach—rooted in policy reform, waste reduction, and *One Health* principles—is essential for safeguarding the planet for future generations.

3.5.5 Conclusion.

Pollution and toxicological threats have emerged as a critical and rapidly intensifying global crisis, exerting far-reaching consequences on biodiversity conservation, ecosystem stability, and human health. The pervasive release of pollutants—ranging from industrial chemicals and heavy metals to plastic waste and air contaminants—has led to widespread environmental degradation, disrupting ecosystem balance and increasing disease burdens across species. These challenges are further exacerbated by man-made climate change, urbanization, and unsustainable industrial and agricultural practices, necessitating urgent and sustained intervention on multiple fronts.

To mitigate these threats, societies must adopt a complex approach that integrates stringent environmental policies, technological innovations, and global cooperation. Strengthening regulatory frameworks to limit emissions, reduce chemical waste, and ban hazardous substances is essential in curbing pollution at its source. Additionally, promoting sustainable industrial and agricultural practices—such as green chemistry, circular economy models, and eco-friendly alternatives—can significantly reduce environmental contamination while fostering economic resilience.

A *One Health* framework, which recognizes the interconnectedness of human, animal, and environmental health, is crucial in addressing the toxicological impacts of pollution. Enhancing environmental monitoring systems, investing in cleaner energy sources, and increasing public awareness through education and advocacy can drive collective action toward pollution reduction. Furthermore, international collaboration is necessary to implement global treaties, such as the *Minamata Convention on Mercury* and the *Stockholm Convention on Persistent Organic Pollutants*, to regulate and eliminate the most hazardous pollutants from the environment.

By prioritizing sustainable development, enforcing evidence-based policies, and fostering cross-sector partnerships, societies can work toward reducing the burden of pollution. Protecting biodiversity and public health from toxicological threats is not only an ecological imperative but also a fundamental requirement for ensuring a healthier and more resilient future for both the planet and its inhabitants.

3.6 Invasive species and biodiversity loss.

3.6.1 Introduction.

Invasive species are one of the leading drivers of global biodiversity loss and pose significant threats to ecosystem stability, food security, and human health. Defined as non-native organisms that establish, spread, and cause ecological or economic harm, invasive species disrupt native ecosystems by outcompeting local species, altering habitat structures, and introducing new pathogens. Human activities such as global trade, travel, man-made climate change, and habitat degradation have accelerated the introduction and spread of invasive species worldwide, exacerbating their impact on biodiversity conservation and public health (Simberloff et al., 2013).

3.6.2 The Impact of Invasive Species on Biodiversity Conservation and Ecosystems.

Invasive species are a major driver of biodiversity conservation decline, causing widespread ecosystem disruption by displacing native species, altering food webs, and transforming ecosystem dynamics. These species, often introduced through global trade, travel, and human activities, thrive in new environments due to a lack of natural predators, competitors, or diseases that would normally regulate their populations. Their impacts are particularly severe in island ecosystems, freshwater environments, and marine habitats, where native species are often poorly adapted to resist invasive threats.

The introduction of invasive predators, herbivores, and plants has led to dramatic shifts in terrestrial ecosystems, often resulting in species extinctions and habitat degradation. One of the most well-documented examples is the introduction of the brown tree snake (*Boiga irregularis*) to Guam. Originally from Papua New Guinea and northern Australia, this arboreal predator was accidentally introduced to the island in the mid-20th century. With no natural predators and abundant prey, the brown tree snake rapidly decimated native bird populations, leading to the near-extinction of several species, including the Guam rail (*Hypotaenidia owstoni*) and the Guam kingfisher (*Todiramphus cinnamominus*) (Savidge, 1987). The loss of these birds disrupted essential ecosystem services such as pollination and seed dispersal, hindering forest regeneration and altering plant community structures.

Similarly, invasive plant species such as Lantana camara and Parthenium hysterophorus have severely impacted native flora by outcompeting indigenous plants, reducing habitat quality, and altering fire regimes. Lantana camara, a rapidly growing shrub, forms dense thickets that inhibit the growth of native vegetation, reduce food availability for herbivores, and increase wildfire frequency by providing highly flammable biomass. Parthenium hysterophorus, an aggressive weed, not only displaces native plants but also releases allelopathic chemicals that inhibit the germination and growth of surrounding vegetation, further degrading ecosystems (Sharma et al., 2005).

Freshwater and marine ecosystems are especially vulnerable to biological invasions, as species introduced through ballast water discharge, aquaculture, and accidental releases often spread rapidly, outcompeting native organisms and altering ecological functions. One of the most notorious aquatic invaders is the zebra mussel (*Dreissena polymorpha*), a Eurasian species that has proliferated throughout North America's freshwater systems. Zebra mussels attach to hard surfaces in massive numbers, clogging water intake systems, damaging infrastructure, and outcompeting native mussels for food and habitat. Their filter-feeding activity also drastically alters nutrient cycling, increasing water clarity but reducing plankton availability, which has cascading effects on fish populations and overall ecosystem health (Higgins & Vander Zanden, 2010).

In marine environments, the lionfish (*Pterois volitans*), native to the Indo-Pacific, has become a devastating invader in the Atlantic Ocean and the Caribbean. Released accidentally through the aquarium trade, lionfish populations have exploded due to their rapid reproduction, venomous spines, and lack of natural predators in their new habitat. They prey aggressively on native fish, leading to significant declines in biodiversity, particularly on coral reefs. Their impact is especially concerning because they reduce populations of herbivorous fish that help control algae growth, indirectly contributing to coral reef degradation (Albins & Hixon, 2008).

Addressing the ecological threats posed by invasive species requires a complex management approach that includes:

1. *Early detection and rapid response* to prevent new introductions and contain outbreaks before they become unmanageable.
2. *Stronger biosecurity measures*, such as stricter regulations on ballast water discharge, plant and animal imports, and quarantine protocols to prevent accidental introductions.
3. *Ecological restoration efforts*, including reintroducing native species, habitat rehabilitation, and biological control programs that utilize natural predators or competitors to suppress invasive populations.
4. *Public awareness and community engagement* to prevent the spread of invasive species through responsible pet ownership, prevention of aquarium releases, and participation in eradication efforts.

Without proactive intervention, invasive species will continue to threaten biodiversity conservation, disrupt ecosystem services, and impose significant economic and environmental costs. A combination of empirically-driven evidence-based public policies, international cooperation, and public involvement is crucial to mitigating their impact and preserving the integrity of native ecosystems.

3.6.3 The Impact of Invasive Species to Human Health.

Invasive species pose significant direct and indirect threats to human health by introducing new pathogens, enhancing disease transmission, and undermining essential resources such as food and water security. These species can alter ecosystems in ways that create new public health risks, often by expanding the range of disease vectors, increasing human exposure to allergens and toxins, and disrupting natural balances that help regulate infectious diseases.

One of the most well-documented examples of invasive species contributing to disease spread is the Asian tiger mosquito (*Aedes albopictus*), a highly adaptable vector that has facilitated the transmission of multiple arboviruses, including dengue, Zika virus, and chikungunya. Originally native to Southeast Asia, *A. albopictus* has expanded its range globally due to increased human travel and trade, particularly through the transport of used tires and other water-holding containers that provide ideal breeding habitats. In regions where these diseases were once rare, the establishment of this mosquito has significantly heightened the risk of outbreaks, posing major public health challenges (Bonizzoni et al., 2013).

Similarly, invasive mammals such as rodents have been implicated in the spread of zoonotic diseases. On many islands, the introduction of invasive rodents, including *Rattus* species, has been linked to an increased prevalence of leptospirosis, a bacterial infection caused by *Leptospira* spp. These rodents contaminate soil and water sources with their urine, creating conditions that facilitate human infection, particularly in tropical and subtropical regions where waterborne transmission is common (Meerburg et al., 2009).

Beyond disease vectors, invasive plant species can also contribute to significant public health burdens. *Parthenium hysterophorus*, an aggressive weed native to the Americas but now widespread in Africa, Asia, and Australia, produces highly allergenic pollen and toxic secondary metabolites. Exposure to these compounds has been associated with severe respiratory conditions, such as asthma and allergic rhinitis, as well as dermatological disorders like contact dermatitis. Agricultural workers and communities living near infested areas are especially vulnerable to these health effects (McFadyen, 1995).

Additionally, invasive microorganisms, including certain cyanobacteria, contribute to the proliferation of harmful algal blooms (HABs), which are becoming more frequent and severe due to climate change and nutrient pollution. These blooms produce potent toxins such as microcystins and saxitoxins, which contaminate drinking water supplies and pose significant risks to human and animal health. Exposure to these toxins can lead to acute poisoning, liver damage, and long-term neurological effects, as seen in numerous incidents of waterborne disease outbreaks (Anderson et al., 2012).

The growing impact of invasive species on human health underscores the need for comprehensive monitoring, early detection, rapid response, and control strategies to mitigate their effects. Effective management requires international collaboration, biosecurity measures, and essential public health functions and services aimed at reducing exposure to invasive species and their associated health risks.

3.6.4 The Impact of Invasive Species on Economic and Ecosystems Costs.

The economic burden of invasive species is staggering, with costs spanning multiple sectors, including agriculture, fisheries, infrastructure, forestry, tourism, and public health. These costs arise not only from direct damages, such as crop loss and ecosystem degradation, but also from the substantial investments required for control, mitigation, and restoration efforts. A global estimate suggests that invasive species inflict economic losses exceeding $1.4 trillion annually—equivalent to nearly 5% of the world's GDP—by disrupting industries that depend on stable ecosystem conditions (Pimentel et al., 2005).

Invasive species pose a significant threat to global food production by reducing crop yields, outcompeting native plant species, and introducing pests and diseases that affect livestock and fisheries. In agriculture, invasive weeds such as *Parthenium hysterophorus* and *Striga hermonthica* reduce productivity by competing for nutrients and water, leading to billions of dollars in crop losses annually. Similarly, invasive insect species like the fall armyworm (*Spodoptera frugiperda*), originally from the Americas, have devastated maize crops across Africa and Asia, forcing governments and farmers to invest heavily in pest management (Early et al., 2016).

Fisheries are also heavily impacted, particularly by invasive aquatic species that disrupt ecosystems and reduce fish populations.

The zebra mussel (*Dreissena polymorpha*), for example, has invaded freshwater systems in North America and Europe, leading to declines in native fish populations by altering food webs and competing with native mollusks. Similarly, the introduction of predatory species like the Nile perch (*Lates niloticus*) in Lake Victoria led to the collapse of native fish stocks, severely affecting local fishing communities and economies reliant on freshwater fisheries (Ogutu-Ohwayo, 1990).

Invasive species also impose substantial costs on infrastructure and maintenance. Invasive tree roots and aquatic plants clog waterways, damage roads, and increase maintenance costs for hydroelectric dams, irrigation systems, and water treatment plants. The spread of invasive aquatic plants such as water hyacinth (*Eichhornia crassipes*) obstructs waterways, impeding transportation and hydropower generation, while zebra mussels and quagga mussels cause extensive damage by clogging water intake pipes and industrial cooling systems, requiring expensive removal and prevention measures (Lovell et al., 2006).

The health sector also bears financial burdens from invasive species, as they contribute to the spread of diseases, increase healthcare costs, and necessitate public health interventions. For example, the invasion of mosquito species such as *Aedes albopictus* and *Aedes aegypti* has increased the prevalence of mosquito-borne diseases, forcing governments to allocate resources toward vector control programs, medical treatments, and disease surveillance. Invasive plants like giant hogweed (*Heracleum mantegazzianum*) produce toxic sap that causes severe skin burns, leading to medical expenses and legal liabilities (Pyšek et al., 2007).

Tourism, another major economic driver, is also significantly affected by invasive species. Degraded natural landscapes, loss of biodiversity, and health risks from invasive organisms deter visitors and reduce revenue for local economies. Coral reef degradation caused by invasive species such as the lionfish (*Pterois volitans*), which preys on native fish populations, has negatively impacted ecotourism and recreational diving industries in the Caribbean and Atlantic (Albins & Hixon, 2008).

Efforts to control invasive species require costly and sustained management strategies. These include chemical treatments such as herbicides and pesticides, biological control measures involving the introduction of natural predators or competitors, and habitat restoration projects designed to recover native biodiversity. Invasive species eradication programs, such as those targeting rats on islands to protect native seabird populations, often require millions of dollars in investment and long-term monitoring to ensure success (Simberloff et al., 2013).

Given the growing economic toll of invasive species, proactive management, early detection, and prevention strategies are crucial to minimizing financial losses. Policies that emphasize biosecurity, public awareness, and ecosystem resilience can help reduce the economic burden while protecting biodiversity and human well-being.

3.6.5 The Impact of Invasive Species on Strategies for Prevention and Management.

Mitigating the impacts of invasive species requires proactive and coordinated strategies at local, national, and international levels. Key approaches include:

1. *Early Detection and Rapid Response:* Monitoring and controlling invasive species before they become established can prevent large-scale ecological damage. Enhanced biosecurity measures, such as stricter border inspections and quarantine protocols, are critical in limiting unintentional introductions (Hulme, 2009).
2. *Ecosystem Restoration and Habitat Management:* Restoring native ecosystems and promoting biodiversity conservation can increase ecosystem resilience to invasions. Native species reintroductions and habitat rehabilitation programs help restore ecosystem balance (D'Antonio & Meyerson, 2002).
3. *Public Awareness and Policy Implementation:* Educating stakeholders, including policymakers, land managers, and the pub-

lic, about the risks of invasive species can lead to improved prevention and management strategies. Strengthening international agreements, such as the *Convention on Biological Diversity (CBD)*, can enhance global cooperation in controlling biological invasions (Pyšek et al., 2020).

4. *Biological Control and Sustainable Management:* Introducing natural predators, parasites, or pathogens can help regulate invasive species populations in an environmentally friendly manner. However, careful risk assessments are necessary to prevent unintended ecosystems consequences (Van Driesche & Bellows, 1996).

3.6.6 Conclusion.

Invasive species represent an escalating global crisis, posing severe threats to biodiversity conservation, ecosystem services stability, human health, and economic sustainability. These species, whether introduced intentionally or accidentally, often lack natural predators or controls in their new environments, allowing them to proliferate unchecked. Their presence disrupts native ecosystems by outcompeting indigenous species, altering food webs, and degrading habitat structures, often leading to local or even global extinctions. The impact extends beyond ecological consequences, as invasive species also introduce novel pathogens, intensify disease transmission, and reduce agricultural productivity, affecting food security and livelihoods.

The increasing interconnectedness of human societies through global trade, travel, and man-made climate change has accelerated the spread of invasive species across continents. Rising global temperatures and shifting environmental conditions create new opportunities for these species to expand their ranges, further exacerbating their ecological and economic impacts. Invasive insects, plants, and pathogens have already caused widespread damage to forests, fisheries, and farmlands, resulting in billions of dollars in annual losses. The agricultural sector, for instance, faces significant threats from invasive pests that destroy crops, necessitating increased pesticide use, which further harms biodiversity conservation and human health. Additionally, invasive aquatic species such as zebra mussels and lionfish disrupt freshwater and marine ecosystems, altering nutrient cycles and depleting native fish populations, impacting local fisheries and economies.

Given the scale and urgency of this issue, a proactive, empirically-driven, evidence-based, and globally coordinated approach is essential. Strengthening biosecurity measures, such as stricter regulations on international trade, improved quarantine protocols, and enhanced border inspections, can help prevent the introduction and spread of invasive species. Investment in early detection and rapid response systems, utilizing advanced surveillance technologies and predictive modeling, is crucial in mitigating their impact before they become unmanageable. Ecosystem restoration efforts, including habitat rehabilitation and native species reintroduction, can improve ecosystem resilience and reduce vulnerability to invasions.

Public engagement and education play a vital role in addressing the invasive species crisis. Encouraging responsible pet ownership, sustainable gardening practices, and public participation in eradication programs can help limit the unintentional spread of invasive species. Policymakers, researchers, conservationists, and local communities must work together to implement adaptive management strategies tailored to regional ecological and socio-economic conditions. Furthermore, fostering international collaboration through treaties, information-sharing networks, and joint conservation initiatives will be crucial in developing long-term solutions to invasive species management.

By prioritizing biodiversity conservation, sustainable land-use practices, and multisectoral partnerships, societies can effectively mitigate the impact of invasive species and build ecosystem resilience. Protecting biodiversity conservation from biological invasions is not only an environmental necessity but also a foundational component of ensuring food security, public health, and economic stability for future generations.

3.7 Human-wildlife conflict and land-use change.

3.7.1 Introduction.

Human-wildlife conflict (HWC) and land-use change are among the most pressing threats to global biodiversity conservation and human well-being. As human populations expand and natural habitats shrink, interactions between humans and wildlife become more frequent, often resulting in negative consequences for both. Land-use changes—driven by agriculture, urbanization, deforestation, and infrastructure development—alter ecosystems, displace species, and disrupt ecosystem balance. These changes not only contribute to biodiversity loss but also exacerbate public health risks by increasing human exposure to zoonotic diseases and reducing access to essential ecosystem services (Dobson et al., 2020).

3.7.2 The Impact of Human-Wildlife Conflict and Land-Use Change on Biodiversity Conservation.

Habitat loss and fragmentation are among the most pressing threats to wildlife, forcing many species into closer contact with human settlements and intensifying competition for space, food, and other resources. As natural habitats shrink due to agricultural expansion, urbanization, and infrastructure development, wildlife increasingly ventures into human-dominated landscapes, leading to conflicts that threaten both biodiversity and livelihoods.

Large carnivores such as lions (*Panthera leo*), tigers (*Panthera tigris*), and wolves (*Canis lupus*) are particularly affected by habitat encroachment. As their natural prey becomes scarce due to habitat degradation or overhunting, these predators often turn to livestock as an alternative food source. This leads to significant economic losses for farmers and herders, who may resort to retaliatory killings, further endangering already vulnerable or endangered species (Treves & Karanth, 2003). In India, for instance, conflicts between tigers and rural communities have escalated as forests are cleared for settlements and agriculture, resulting in increased tiger poaching despite conservation efforts.

Herbivorous species also contribute to human-wildlife conflicts, particularly when they raid crops or damage infrastructure. Elephants (*Loxodonta africana* and *Elephas maximus*), for example, frequently destroy farmlands in Africa and Asia, leading to significant financial losses for subsistence farmers. Similarly, wild boars (*Sus scrofa*) and deer species damage crops, prompting lethal control measures such as culling and fencing, which can have unintended ecological consequences (Goswami et al., 2014). In some cases, retaliatory killings of herbivores also affect their predators, as reduced prey availability forces carnivores to seek alternative food sources, sometimes including domestic animals or even humans.

Beyond direct human-wildlife conflict, habitat fragmentation severely disrupts the natural behaviors and survival strategies of numerous species. Many migratory species, including ungulates like wildebeest (*Connochaetes taurinus*) and pronghorns (*Antilocapra americana*), rely on vast, uninterrupted landscapes to complete seasonal movements. However, expanding roads, fences, and urban development increasingly block these migration corridors, leading to population declines due to restricted access to food and water resources.

For species with specific breeding and nesting requirements, habitat fragmentation can be particularly devastating. Amphibians, for example, are highly dependent on wetlands for reproduction, but widespread drainage of wetlands and deforestation have decimated populations by eliminating breeding sites. Similarly, road networks and urban sprawl have become deadly barriers for slow-moving or territorial species such as koalas (*Phascolarctos cinereus*) and jaguars (*Panthera onca*), leading to increased roadkill incidents and genetic isolation (Laurance et al., 2009).

When populations become isolated due to habitat fragmentation, genetic exchange between groups decreases, making species more vulnerable to inbreeding and genetic bottlenecks. This reduces genetic diversity, which is critical for adaptive resilience to environmental changes, disease, and other threats. For example, studies on Florida panthers (*Puma concolor coryi*) have shown that habitat fragmentation has led to severe inbreeding, resulting in genetic defects such as heart problems and kinked tails (Hedrick et al., 2010). Conservationists have attempted genetic rescue by introducing individuals from other populations, but such interventions are costly and logistically challenging.

The consequences of habitat loss and fragmentation extend beyond individual species, triggering cascading effects throughout ecosystems. Disruptions in predator-prey dynamics can alter entire food webs, leading to imbalances that threaten biodiversity resilience. For instance, the decline of large predators due to habitat fragmentation often results in mesopredator release, where medium-sized predators like coyotes (*Canis latrans*) or raccoons (*Procyon lotor*) proliferate unchecked, leading to declines in smaller prey species (Prugh et al., 2009). Similarly, the loss of keystone herbivores, such as elephants, which play a crucial role in shaping landscapes, can alter vegetation structure and decrease biodiversity conservation.

In marine ecosystems, habitat degradation from coastal development, pollution, and man-made climate change has had similar cascading effects. Coral reef destruction, for example, reduces habitat complexity and shelter for fish species, ultimately disrupting entire reef ecosystems (Hughes et al., 2017). In the Arctic, declining sea ice has forced polar bears (*Ursus maritimus*) to spend more time on land, increasing their interactions with human settlements and leading to increased human-wildlife conflicts.

Addressing the impacts of habitat loss and fragmentation requires proactive conservation strategies, including the establishment of wildlife corridors, protected areas, and sustainable land-use policies. Landscape connectivity projects, such as the *Yellowstone-to-Yukon Conservation Initiative* in North America, aim to link fragmented habitats, allowing species to move freely and maintain genetic diversity (Carroll et al., 2018). In Africa, community-based conservation programs that involve local populations in wildlife protection efforts have shown promise in reducing human-wildlife conflicts while promoting sustainable coexistence.

Moreover, advancements in technology, such as GPS tracking, remote sensing, and AI-powered monitoring, are improving conservationists' ability to track wildlife movements and identify high-risk conflict zones. Governments and conservation organizations are also increasingly investing in compensation schemes for farmers affected by wildlife conflicts, incentivizing coexistence rather than retaliation.

Ultimately, balancing human development with biodiversity conservation is crucial to ensuring the long-term survival of wildlife and the ecosystem services they provide. Without strategic interventions, the accelerating pace of habitat destruction and fragmentation will continue to threaten global biodiversity conservation, with irreversible consequences for both wildlife and human communities.

3.7.3 The Impact of Human-Wildlife Conflict and Land-Use Change on Human Health.

Human-wildlife conflict (HWC) and land-use changes significantly heighten public health risks by creating pathways for zoonotic disease spillover. The destruction and fragmentation of natural habitats force wildlife into closer proximity with human populations, increasing the likelihood of pathogen transmission. Deforestation, agricultural expansion, urbanization, and wildlife trade are key drivers of emerging infectious diseases (EIDs), as they alter ecosystem balances and provide new opportunities for viruses, bacteria, and parasites to jump from animals to humans.

Extensive research has linked deforestation and agricultural expansion to major outbreaks of zoonotic diseases, including Ebola, Nipah virus, and malaria. As forests are cleared for plantations, livestock grazing, and human settlements, disease-carrying wildlife is displaced and forced into closer contact with people. For example, Ebola virus outbreaks in Central and West Africa have often occurred near deforested regions, where human encroachment increases interactions with fruit bats, the virus's natural reservoir (Plowright et al., 2017).

Similarly, the clearing of forests for pig farming in Malaysia played a direct role in the emergence of Nipah virus in 1998. Fruit bats, deprived of their natural food sources, sought out orchards near pig farms, leading to viral transmission to livestock and, ultimately, to humans through close contact with infected pigs (Daszak et al., 2006). This pattern of habitat destruction driving pathogen spillover is not limited to viral diseases—vector-borne diseases such as malaria are also exacerbated by land-use changes.

Forest fragmentation creates ideal conditions for the spread of disease-carrying vectors such as mosquitoes, ticks, and rodents, which thrive in disturbed environments. Research has shown a direct correlation between deforestation in the Amazon and increased malaria transmission. The clearing of trees creates stagnant water pools that serve as breeding sites for *Anopheles* mosquitoes, the primary vectors of malaria, leading to surges in infection rates among local populations (Vittor et al., 2006).

Similarly, Lyme disease, a bacterial infection transmitted by ticks, has expanded in North America due to habitat fragmentation. As large predators decline and forest edges proliferate, populations of white-footed mice (*Peromyscus leucopus*)—key reservoirs for Lyme disease—have exploded, increasing tick densities and the likelihood of human exposure (Allan et al., 2003). These examples underscore how ecosystem disturbances directly shape the epidemiology of infectious diseases, making land conservation an essential public health strategy.

Beyond deforestation and land conversion, the wildlife trade and human encroachment into previously undisturbed ecosystems further amplify the risks of disease emergence. The COVID-19 pandemic, caused by SARS-CoV-2, highlighted the dangers of wildlife markets and habitat destruction in facilitating novel infectious diseases (Jones et al., 2008). The close confinement of diverse animal species in live markets creates ideal conditions for viral recombination and spillover to humans.

SARS-CoV-1, which caused the 2002–2003 SARS outbreak, is believed to have originated in bats before jumping to civet cats and humans through wildlife markets in China. Similarly, Middle East Respiratory Syndrome (MERS), first identified in 2012, emerged from camel-to-human transmission, with the virus likely originating in bats (Cui et al., 2019). As humans continue to encroach on wildlife habitats for food, medicine, and trade, the risk of encountering novel pathogens increases, heightening the potential for future pandemics.

Given the clear link between land-use change and EIDs, sustainable land management and biodiversity conservation must be prioritized as critical public health strategies. Protecting intact ecosystems serves as a natural barrier against disease spillover by maintaining ecosystem balances and limiting human-wildlife interactions. Strategies such as reforestation, habitat corridors, and sustainable agricultural practices can help mitigate the risk of zoonotic disease transmission while promoting biodiversity resilience.

Additionally, strengthening regulations on wildlife trade and improving disease surveillance in high-risk areas are essential components of pandemic prevention. Integrated approaches such as the *One Health* framework, which recognizes the inextricable interconnectedness of human, animal, and environmental health, provide a holistic and comprehensive strategy for addressing EIDs (Karesh et al., 2012).

As global population growth and economic development drive further land-use changes, proactive biodiversity conservation and essential public health services and policies will be necessary to prevent future pandemics and protect both human and environmental well-being.

3.7.4 Economic and Social Impacts of Human-Wildlife Conflict and Land-Use Change.

Human-wildlife conflict (HWC) and land-use changes impose severe economic burdens on communities that rely on agriculture, tourism, and natural resources. These disruptions not only lead to direct financial losses but also contribute to long-term economic instability, affecting livelihoods, food security, and regional development. The economic consequences of HWC and habitat transformation extend across multiple sectors, influencing rural economies, global markets, and ecosystem services.

Crop destruction and livestock predation by wildlife result in substantial economic losses for farmers, particularly in regions where agriculture is the primary source of income. Large herbivores such as elephants (*Loxodonta africana*) and wild boars (*Sus scrofa*) often raid farmlands, consuming and trampling crops, sometimes leading to complete harvest failures. Studies estimate that elephant crop raids in parts of Africa and Asia can result in losses of up to 15% of household annual income (Naughton-Treves et al., 2005).

Similarly, carnivore predation on livestock disrupts rural economies by reducing income streams and increasing expenditures on protective measures such as fencing, night enclosures, and compensation schemes. For example, wolf (*Canis lupus*) attacks on livestock in Europe and North America have forced farmers to invest heavily in deterrent strategies or abandon traditional pastoral livelihoods altogether (Dickman et al., 2011). In extreme cases, retaliatory killings of predators, such as lions (*Panthera leo*) and tigers (*Panthera tigris*), lead to further ecological imbalances and conservation challenges.

HWC and habitat degradation also affect the tourism industry, which depends on intact ecosystems and thriving wildlife populations. In many countries, wildlife tourism generates significant revenue, supporting biodiversity conservation programs and providing employment opportunities. However, when conflicts arise, local communities often perceive wildlife as a threat rather than an asset, leading to reduced support for conservation initiatives and an increase in poaching or habitat destruction.

For instance, in regions where safari tourism is a major economic driver, elephant and lion poaching have escalated due to growing animosity between local communities and conservation authorities (Lindsey et al., 2013). Furthermore, infrastructure development such as roads, dams, and urban expansion fragments natural habitats, diminishing the appeal of ecotourism destinations and reducing visitor numbers. Declines in wildlife tourism result in financial losses for businesses, governments, and conservation projects that rely on entrance fees and international funding.

In many developing nations, land is a critical resource for subsistence farming, grazing, and settlement. As land-use changes encroach on wildlife habitats, competition for space intensifies, often leading to violent conflicts between communities, conservationists, and governments. In some cases, disputes over land use have escalated into armed conflicts, as seen in regions where indigenous groups and pastoralists are forcibly displaced to make way for commercial agriculture or conservation zones (Benjaminsen et al., 2012).

Additionally, human displacement due to conservation policies—such as the establishment of protected areas—can generate long-term socio-economic instability. Communities that depend on forests and wetlands for food, medicine, and cultural practices may lose access to these resources, leading to increased poverty and resentment toward conservation efforts. Such conflicts can undermine biodiversity protection strategies, as affected populations may engage in illegal hunting or logging as alternative sources of income.

Beyond direct economic losses, land-use change and biodiversity conservation degradation weaken essential ecosystem services that sustain local and global economies. Forests, wetlands, and grasslands provide critical benefits such as water purification, carbon sequestration, pollination, and climate regulation. The destruction of these ecosystems reduces their capacity to support agriculture, fisheries, and clean water supplies, further exacerbating economic vulnerabilities.

For example, the decline of pollinators due to habitat loss has had measurable economic impacts on global food production. Studies estimate that the loss of bees and other pollinators could result in annual agricultural losses of up to $577 billion worldwide (IPBES, 2016). Similarly, deforestation-driven disruptions in hydrological cycles contribute to water scarcity and reduced hydroelectric power generation, increasing energy costs and limiting economic development.

Addressing the economic consequences of HWC and land-use change requires integrated and holistic strategies that balance development with biodiversity conservation. Sustainable land-use planning, including wildlife corridors, agroforestry, and participatory conservation programs, can help reduce conflicts while maintaining ecosystem integrity. Additionally, economic incentives such as ecotourism revenue-sharing, compensation funds for farmers, and investment in alternative livelihoods can mitigate financial losses and promote coexistence between human communities and wildlife.

Strengthening policies that protect both biodiversity conservation and local economies is crucial for long-term sustainability. Governments, conservation organizations, and private stakeholders must collaborate to develop resilient strategies that minimize economic risks while fostering ecosystem and environmental stewardship. By prioritizing coexistence solutions, societies can ensure that economic growth does not come at the expense of environmental and social stability.

3.7.5 Strategies for Mitigation and Management of Human-Wildlife Conflict and Land-Use Change.

Addressing HWC and mitigating the effects of land-use change require an integrated, multi-disciplinary approach that balances biodiversity conservation efforts with human development. Key strategies include:

1. *Habitat Protection and Restoration:* Expanding protected areas, creating wildlife corridors, and restoring degraded ecosystems can reduce habitat fragmentation and minimize human-wildlife interactions (Newmark, 2008).
2. *Sustainable Land-Use Planning:* Implementing agroforestry, conservation agriculture, and land zoning policies can help balance

human needs with ecological integrity, reducing conflicts over space and resources (Scherr & McNeely, 2008).

3. *Community-Based Conflict Mitigation:* Engaging local communities in wildlife biodiversity conservation through incentive programs, compensation structures, and participatory decision-making fosters coexistence and reduces retaliatory killings (Zimmermann et al., 2009).

4. *Strengthening Disease Surveillance and Public Health Systems:* Monitoring emerging zoonotic diseases, regulating wildlife trade, and enhancing healthcare infrastructure are critical strategies for mitigating the health risks linked to land-use change and human-wildlife conflict (Daszak et al., 2020). In addition, strengthening the capacity for essential public health functions and services at all socio-ecological levels is foundational for sustaining evidence-based practices. This approach is essential for reinforcing global disease surveillance systems and ensuring effective public health responses worldwide.

5. *International Collaboration and Policy Integration:* Global initiatives such as the Convention on Biological Diversity (CBD) and the *One Health* framework emphasize the need for cross-sectoral cooperation to address biodiversity loss, zoonotic diseases, and land-use conflicts (Karesh et al., 2012).

3.7.6 Conclusion.

Human-wildlife conflict (HWC) and land-use change are rapidly accelerating the loss of biodiversity conservation, posing significant and growing threats to both ecosystems and the public's health. As human populations expand and agricultural, industrial, and urban development continue to encroach on natural habitats, ecosystems are being disrupted, species are pushed toward extinction, and ecosystem services stability is compromised. This encroachment increases human-wildlife interactions, often leading to conflicts that negatively impact both wildlife and human communities. Additionally, habitat fragmentation and alteration create conditions that facilitate the spread of zoonotic diseases, with wildlife acting as intermediaries in the transmission of pathogens from animals to humans. This growing overlap between human and wildlife territories increases the risk of future pandemics and emerging diseases, further exacerbating public health threats.

To mitigate these interconnected threats, urgent and coordinated action is required across multiple sectors. First, promoting sustainable land-use practices that balance human development needs with the preservation of natural habitats is critical. This includes adopting sustainable subsistence agriculture, forestry, and urban planning practices that reduce habitat destruction, fragmentation, and pollution. Efforts to restore degraded ecosystems and establish wildlife corridors can help reconnect fragmented habitats, allowing species to migrate, breed, and maintain genetic diversity, thereby reducing extinction risks and enhancing ecosystem resilience.

Strengthening biodiversity conservation efforts is also paramount. Protecting biodiversity through the establishment and expansion of protected areas, alongside the implementation of community-based conservation initiatives, can safeguard critical ecosystems and species. Furthermore, addressing human-wildlife conflict through non-lethal methods—such as improved fencing, deterrents, and compensation schemes—can reduce the negative impacts of wildlife on agricultural and pastoral activities, while fostering coexistence between people and wildlife.

Integrating essential public health services and environmental policies is essential in addressing the complex challenges posed by HWC and land-use change. By adopting a *One Health* approach, which recognizes the interconnectedness of human, animal, and environmental health, policymakers can develop more effective strategies to prevent disease transmission, mitigate human-wildlife conflict, and conserve biodiversity. Strengthening disease surveillance and early response systems, regulating wildlife trade, and improving healthcare infrastructure in regions where land-use changes and

HWC are most prevalent can reduce the public health risks associated with these environmental disruptions.

Fostering coexistence between humans and wildlife requires collaboration at local, national, and global levels. Governments, conservation organizations, local communities, and businesses must work together to develop and implement policies and practices that prioritize the long-term health of ecosystems and human populations alike. By embedding sustainable practices, fostering biodiversity conservation, and addressing public health risks, societies can protect the natural world while ensuring the economic and social well-being of future generations. In doing so, we can safeguard biodiversity, mitigate the risks of emerging diseases, and promote long-term ecological and public health security.

3.8 References.

1. Allan, B. F., Keesing, F., & Ostfeld, R. S. (2003). Effect of forest fragmentation on Lyme disease risk. *Conservation Biology, 17*(1), 267-272.
2. Albins, M. A., & Hixon, M. A. (2008). Invasive Indo-Pacific lionfish *Pterois volitans* reduce recruitment of Atlantic coral-reef fishes. *Marine Ecology Progress Series, 367*, 233-238.
3. Anderson, D. M., et al. (2012). Harmful algal blooms and eutrophication: Nutrient sources, composition, and consequences. *Estuaries and Coasts, 35*(3), 725-735.
4. Benjaminsen, T. A., Maganga, F. P., & Abdallah, J. M. (2012). The Kilosa killings: Political ecology of a farmer–herder conflict in Tanzania. *Development and Change, 43*(5), 1141-1169.
5. Bogich, T. L., et al. (2012). Global patterns of zoonotic disease in mammals. *Trends in Parasitology, 28*(3), 90-96.
6. Bonizzoni, M., et al. (2013). The invasive mosquito species *Aedes albopictus*: current knowledge and future perspectives. *Trends in Parasitology, 29*(9), 460-468.

7. Borrelle, S. B., et al. (2020). Predicted growth in plastic waste exceeds efforts to mitigate plastic pollution. *Science*, 369(6510), 1515-1518.

8. Carroll, C., Noss, R. F., & Paquet, P. C. (2018). Rediscovering landscape connectivity: Challenges and opportunities in the 21st century. *Conservation Biology*, 32(4), 910-921.

9. Ceballos, G., Ehrlich, P. R., & Dirzo, R. (2015). Biological annihilation via the ongoing sixth mass extinction signaled by vertebrate population losses and declines. *Proceedings of the National Academy of Sciences, 114*(30), E6089-E6096.

10. Chazdon, R. L. (2008). Beyond deforestation: restoring forests and ecosystem services on degraded lands. *Science, 320*(5882), 1458-1460.

11. Chen, I. C., et al. (2011). Rapid range shifts of species associated with high levels of climate warming. *Science, 333*(6045), 1024-1026.

12. Cui, J., Li, F., & Shi, Z. L. (2019). Origin and evolution of pathogenic coronaviruses. *Nature Reviews Microbiology, 17*(3), 181-192.

13. D'Antonio, C., & Meyerson, L. A. (2002). Exotic plant species as problems and solutions in ecological restoration: A synthesis. *Restoration Ecology*, 10(4), 703-713.

14. Daszak, P., et al. (2000). Emerging infectious diseases of wildlife— threats to biodiversity and human health. *Science*, 287(5452), 443-449.

15. Daszak, P., et al. (2001). Anthropogenic environmental change and the emergence of infectious diseases in wildlife. *Acta Tropica*, 78(2), 103-116.

16. Daszak, P., Cunningham, A. A., & Hyatt, A. D. (2006). Emerging infectious diseases of wildlife—Threats to biodiversity and human health. *Science, 287*(5452), 443-449.

17. Daszak, P., et al. (2020). Ecology and economics of pandemics. *Science*, 368(6496), 379-382.

18. Desforges, J. P., et al. (2018). Predicting global killer whale population collapse from PCB pollution. *Science*, 361(6409), 1373-1376.

19. Destoumieux-Garzón, D., Mavingui, P., Boetsch, G., et al. (2018). The One Health concept: 10 years old and a long road ahead. *Frontiers in Veterinary Science, 5*, 14.

20. Díaz, S., Settele, J., Brondízio, E. S., et al. (2019). Pervasive human-driven decline of life on Earth points to the need for transformative change. *Science, 366*(6471), eaax3100.

21. Dickman, A. J., et al. (2011). The costs of coexistence: Understanding human-wildlife conflict and conservation. *Animal Conservation*, 14(6), 502-503.

22. Dickman, A. J., Macdonald, E. A., & Macdonald, D. W. (2011). A review of financial instruments to pay for predator conservation and encourage human-carnivore coexistence. *Proceedings of the National Academy of Sciences, 108*(34), 13937-13944.

23. Dobson, A. P., et al. (2020). Biodiversity loss and emerging infectious disease: An urgent need to adopt a One Health approach. *Nature Sustainability*, 3(10), 677-681.

24. Dobson, A. P., et al. (2020). Ecology and economics for pandemic prevention. *Science, 369*(6502), 379-381.

25. Early, R., Bradley, B. A., Dukes, J. S., et al. (2016). Global threats from invasive alien species in the twenty-first century and national response capacities. *Nature Communications, 7*, 12485.

26. Foley, J. A., et al. (2005). Global consequences of land use. *Science, 309*(5734), 570-574.

27. Frick, W. F., et al. (2016). Disease alters macroecological patterns of North American bats. *Ecology Letters*, 19(3), 320-329.

28. Frick, W. F., Pollock, J. F., Hicks, A. C., et al. (2016). An emerging disease causes regional population collapse of a common North American bat species. *Science, 329*(5992), 679-682.

29. Gao, R., Cao, B., Hu, Y., et al. (2013). Human infection with a novel avian-origin influenza A (H7N9) virus. *New England Journal of Medicine, 368*(20), 1888-1897.

30. Gibb, R., et al. (2020). Zoonotic host diversity increases in human-dominated ecosystems. *Nature, 584*(7821), 398-402.

31. Gibson, L., et al. (2011). Primary forests are irreplaceable for sustaining tropical biodiversity. *Nature, 478*(7369), 378-381.

32. Gore, A. C., et al. (2015). Endocrine-disrupting chemicals: Current understanding and future directions. *Nature Reviews Endocrinology*, 11(11), 653-660.

33. Gore, A. C., et al. (2015). "EDC-2: The Endocrine Society's Second Scientific Statement on Endocrine-Disrupting Chemicals." *Endocrine Reviews*, 36(6), E1–E150.

34. Goswami, V. R., et al. (2014). Tigers teach us how to live with large carnivores. *Scientific Reports*, 4, 5673.

35. Goswami, V. R., Medhi, K., Nichols, J. D., & Oli, M. K. (2014). Mechanistic understanding of human-wildlife conflict through a novel application of dynamic occupancy models. *Conservation Biology*, 28(6), 1687-1695.

36. Grandjean, P., & Landrigan, P. J. (2014). Neurobehavioral effects of developmental toxicity. *The Lancet Neurology*, 13(3), 330-338.

37. Haddad, N. M., et al. (2015). Habitat fragmentation and its lasting impact on Earth's ecosystems. *Science Advances, 1*(2), e1500052.

38. Hedrick, P. W., Fredrickson, R. J., & Kalinowski, S. T. (2010). Genetic rescue and inbreeding depression in the Florida panther. *Animal Conservation*, 13(4), 364-373.

39. Higgins, S. N., & Vander Zanden, M. J. (2010). "What a difference a species makes: A meta–analysis of zebra mussel impacts on freshwater ecosystems." *Ecological Monographs*, 80(2), 179-196.

40. Hoegh-Guldberg, O., Jacob, D., Taylor, M., et al. (2017). Impacts of 1.5°C global warming on natural and human systems. *IPCC Special Report on Global Warming of 1.5°C*, 175-311.

41. Hughes, T. P., et al. (2017). Coral reefs in the Anthropocene. *Nature, 546*(7656), 82-90.

42. Hughes, T. P., Kerry, J. T., Álvarez-Noriega, M., et al. (2017). Global warming and recurrent mass bleaching of corals. *Nature*, 543(7645), 373-377.

43. Hulme, P. E. (2009). Trade, transport and trouble: managing invasive species pathways in an era of globalization. *Journal of Applied Ecology*, 46(1), 10-18.

44. Intergovernmental Panel on Climate Change (IPCC). (2021). *Climate Change 2021: The Physical Science Basis*. Intergovernmental Panel on Climate Change.

45. Intergovernmental Science-Policy Platform on Biodiversity and Ecosystem Services (IPBES). (2016). *The assessment report on pollinators, pollination and food production.*

46. Intergovernmental Science-Policy Platform on Biodiversity and Ecosystem Services (IPBES). (2019). *Global Assessment Report on Biodiversity and Ecosystem Services*. IPBES Secretariat.

47. Intergovernmental Science-Policy Platform on Biodiversity and Ecosystem Services (IPBES). (2020). *Workshop Report on Biodiversity and Pandemics*. Intergovernmental Science-Policy Platform on Biodiversity and Ecosystem Services.

48. Jones, K. E., Patel, N. G., Levy, M. A., et al. (2008). Global trends in emerging infectious diseases. *Nature, 451*(7181), 990-993.

49. Karesh, W. B., Dobson, A., Lloyd-Smith, J. O., et al. (2012). Ecology of zoonoses: Natural and unnatural histories. *The Lancet, 380*(9857), 1936-1945.

50. Keesing, F., Belden, L. K., Daszak, P., Dobson, A., Harvell, C. D., Holt, R. D., ... & Ostfeld, R. S. (2010). Impacts of biodiversity on the emergence and transmission of infectious diseases. *Nature, 468*(7324), 647-652.

51. Lambin, E. F., et al. (2018). The role of supply-chain initiatives in reducing deforestation. *Nature Climate Change, 8*(2), 109-116.

52. Landrigan, P. J., et al. (2018). The Lancet Commission on pollution and health. *The Lancet*, 391(10119), 462-512.

53. Laurance, W. F., Goosem, M., & Laurance, S. G. (2009). Impacts of roads and linear clearings on tropical forests. *Trends in Ecology & Evolution*, 24(12), 659-669.

54. Lelieveld, J., et al. (2020). Loss of life expectancy from air pollution compared to other risk factors: a worldwide perspective. *Cardiovascular Research*, 116(11), 1910-1917.

55. Lelieveld, J., et al. (2020). "Effects of fossil fuel and total anthropogenic emission removal on public health and climate." *PNAS*, 117(15), 9124–9130.

56. Lindsey, P. A., Alexander, R., Mills, M. G. L., et al. (2013). The impact of hunting on African lion (*Panthera leo*) and implications for its conservation in the wild. *Biological Conservation, 134*(4), 548-558.

57. Lovell, S. J., Stone, S. F., & Fernandez, L. (2006). The economic impacts of aquatic invasive species: A review of the literature. *Agricultural and Resource Economics Review, 35*(1), 195-208.

58. Malhi, Y., et al. (2008). Climate change, deforestation, and the fate of the Amazon. *Science, 319*(5860), 169-172.

59. McFadyen, R. E. (1995). Parthenium weed and human health in Queensland. *Australian Family Physician*, 24(8), 1455-1459.

60. McMichael, A. J. (2015). Climate change and global health: Present and future risks. *The Lancet, 386*(10006), 1861-1866.

61. Meerburg, B. G., et al. (2009). Rodent-borne diseases and their risks for public health. *Critical Reviews in Microbiology*, 35(3), 221-270.

62. Messina, J. P., et al. (2019). The current and future global distribution and population at risk of dengue. *Nature Microbiology, 4*(9), 1508-1515.

63. Millennium Ecosystem Assessment. (2005). *Ecosystems and Human Well-being: Biodiversity Synthesis*. World Resources Institute.

64. Myers, S. S., Gaffikin, L., Golden, C. D., et al. (2013). Human health impacts of ecosystem alteration. *Proceedings of the National Academy of Sciences, 110*(47), 18753-18760.

65. Myers, S. S., et al. (2017). Climate change and global food systems: Potential impacts on food security and undernutrition. *Annual Review of Public Health, 38*(1), 259-277.

66. Naughton-Treves, L., Treves, A., Chapman, C., & Wrangham, R. (2005). Temporal patterns of crop-raiding by primates: Linking food availability in croplands and adjacent forest. *Journal of Applied Ecology, 36*(4), 699-710.
67. Newbold, T., et al. (2016). Has land use pushed terrestrial biodiversity beyond the planetary boundary? *Science, 353*(6296), 288-291.
68. Newmark, W. D. (2008). Isolation of African protected areas. *Frontiers in Ecology and the Environment, 6*(6), 321-328.
69. Ogden, N. H., et al. (2014). Climate change and the expansion of Lyme disease risk areas in Canada. *Canadian Medical Association Journal, 186*(10), E320-E326.
70. Ogden, N. H., Radojevic, M., Caminade, C., & Gachon, P. (2014). Climate change and the risk of emerging vector-borne diseases in Canada. *Canadian Communicable Disease Report, 40*(9), 209-218.
71. Ogutu-Ohwayo, R. (1990). The decline of the native fishes of Lake Victoria and Kyoga (East Africa) and the impact of introduced species, especially the Nile perch, *Lates niloticus*, and the Nile tilapia, *Oreochromis niloticus*. *Environmental Biology of Fishes, 27*, 81-96.
72. Patz, J. A., et al. (2005). Impact of regional climate change on human health. *Nature, 438*(7066), 310-317.
73. Pecl, G. T., Araújo, M. B., Bell, J. D., et al. (2017). Biodiversity redistribution under climate change: Impacts on ecosystems and human well-being. *Science, 355*(6332), eaai9214.
74. Piarroux, R., et al. (2011). Understanding the cholera epidemic, Haiti. *Emerging Infectious Diseases, 17*(7), 1161-1168.
75. Piarroux, R., Faucher, B., Cholera, H., & Outbreak, R. (2011). Cholera in Haiti: The equity agenda and the future of tropical epidemiology. *The American Journal of Tropical Medicine and Hygiene, 84*(1), 15-17.
76. Pimentel, D., et al. (2005). Economic and environmental threats of alien plant, animal, and microbe invasions. *Ecological Economics*, 52(3), 273-288.

77. Plowright, R. K., et al. (2017). Pathways to zoonotic spillover. *Nature Reviews Microbiology*, 15(8), 502-510.

78. Plowright, R. K., Foley, P., Field, H. E., et al. (2017). Urbanization, land-use change, and the ecology of infectious diseases. *Trends in Ecology & Evolution, 32*(2), 104-116.

79. Poloczanska, E. S., et al. (2013). Global imprint of climate change on marine life. *Nature Climate Change, 3*(10), 919-925.

80. Prugh, L. R., Stoner, C. J., Epps, C. W., et al. (2009). The rise of the mesopredator. BioScience, 59(9), 779-791.

81. Pyšek, P., et al. (2020). Scientists' warning on invasive alien species. *Biological Reviews*, 95(6), 1511-1534.

82. Pyšek, P., Richardson, D. M., Rejmánek, M., et al. (2007). Alien plants in checklists and floras: Towards better communication between taxonomists and ecologists. *Biodiversity and Conservation, 13*(2), 297-308.

83. Ravenscroft, P., et al. (2009). *Arsenic Pollution: A Global Synthesis.* Wiley-Blackwell.

84. Rochman, C. M., et al. (2016). Anthropogenic debris in seafood: Plastic debris and fibers from textiles in fish and bivalves sold for human consumption. *Scientific Reports*, 5, 14340.

85. Rocklöv, J., & Dubrow, R. (2020). Climate change: An enduring challenge for vector-borne disease prevention and control. *Nature Immunology, 21*(5), 479-483.

86. Ryan, S. J., Carlson, C. J., Mordecai, E. A., & Johnson, L. R. (2019). Global expansion and redistribution of Aedes-borne virus transmission risk with climate change. *PLoS Neglected Tropical Diseases, 13*(3), e0007213.

87. Ryan, S. J., et al. (2019). Climate change and vector-borne diseases: emerging trends and future threats. *The Lancet Planetary Health*, 3(10), e360-e369.

88. Savidge, J. A. (1987). Extinction of an island forest avifauna by an introduced snake. *Ecology*, 68(3), 660-668.

89. Scheele, B. C., Pasmans, F., Skerratt, L. F., et al. (2019). Amphibian fungal panzootic causes catastrophic and ongoing loss of biodiversity. *Science, 363*(6434), 1459-1463.

90. Scherr, S. J., & McNeely, J. A. (2008). Biodiversity conservation and agricultural sustainability: Towards a new paradigm of 'ecoagriculture' landscapes. *Philosophical Transactions of the Royal Society B: Biological Sciences*, 363(1491), 477-494.

91. Sharma, G. P., et al. (2005). Invasive species: Taxonomic trends and management issues. *Current Science*, 89(8), 1325-1331.

92. Sharma, G. P., Raghubanshi, A. S., & Singh, J. S. (2005). "Lantana invasion: An overview." *Weed Biology and Management*, 5(4), 157–165.

93. Simberloff, D., Martin, J. L., Genovesi, P., et al. (2013). Impacts of biological invasions: what's what and the way forward. *Trends in Ecology & Evolution, 28*(1), 58-66.

94. Smith, V. H., & Schindler, D. W. (2009). Eutrophication science: where do we go from here? *Trends in Ecology & Evolution*, 24(4), 201-207.

95. Treves, A., & Karanth, K. U. (2003). Human-carnivore conflict and perspectives on carnivore management worldwide. *Conservation Biology*, 17(6), 1491-1499.

96. Van Driesche, R. G., & Bellows, T. S. (1996). *Biological control.* Springer Science & Business Media.

97. Vicedo-Cabrera, A. M., Scovronick, N., Sera, F., et al. (2021). The burden of heat-related mortality attributable to recent human-induced climate change. *Nature Climate Change, 11*(6), 492-500.

98. Vittor, A. Y., Gilman, R. H., Tielsch, J., et al. (2006). The effect of deforestation on the human-biting rate of *Anopheles darlingi*, the primary vector of malaria in the Peruvian Amazon. *The American Journal of Tropical Medicine and Hygiene, 74*(1), 3-11.

99. Walker, W., et al. (2020). The role of forest conversion, degradation, and disturbance in the carbon dynamics of Amazon indigenous territories and protected areas. *Proceedings of the National Academy of Sciences, 117*(6), 3015-3025.

100. Wood, T. J., & Goulson, D. (2017). The environmental risks of neonicotinoid pesticides: a review of the evidence post-2013. *Environmental Science and Pollution Research*, 24(21), 17285-17325.
101. Woolhouse, M. E. J., & Gowtage-Sequeria, S. (2005). Host range and emerging and reemerging pathogens. *Emerging Infectious Diseases*, 11(12), 1842-1847.
102. Wu, F., et al. (2020). A new coronavirus associated with human respiratory disease in China. *Nature*, 579(7798), 265-269.
103. Zimmermann, A., et al. (2009). Can conservation incentives reduce conflicts? *Oryx*, 43(3), 295-304.

4.0

Environmental Health.

Introduction to Environmental Health

Environmental Health is a branch of public health that focuses on the interactions between people and their surroundings, aiming to prevent disease and promote wellness and well-being. It encompasses a wide range of factors, including air and water quality, food safety, hazardous waste management, occupational health, and Anthropocene (i.e., man-made) climate change. The field is essential in mitigating environmental risks such as pollution, chemical exposure, and infectious disease outbreaks, all of which can significantly impact human health. By integrating scientific research, education, practice, policy development, and community engagement, environmental health professionals work to create safer, healthier environments for individuals and populations (CDC, 2023; NIEHS, 2023; WHO, 2023).

4.1 Environmental Health: What is it and Why is it Important in Conservation Medicine?

4.1.1 Introduction.

Environmental Health plays a pivotal role in *Conservation Medicine* by examining the intricate relationships between human, animal, and ecosystem health. This multidisciplinary and multisectoral field acknowledges that the well-being of individuals and populations is directly shaped by environmental conditions, highlighting the necessity of integrated, holistic, and comprehensive approaches to disease surveillance, early response, and prevention. Additionally, it emphasizes the management of ecosystem services and biodiversity conservation as essential strategies for sustaining health across all species.

As human activity continues to alter natural habitats, disrupt ecosystem balance, and contribute to Anthropocene (i.e., manmade) climate change, the interconnectedness between ecosystems and human health has become increasingly evident. Emerging infectious diseases (EIDs), many of which originate at the human-animal-environment interface, exemplify the urgent need for a holistic, *One Health* approach to mitigate public health risks. Environmental degradation and biodiversity loss further compound these challenges, weakening ecosystems' natural defenses and increasing vulnerability to disease outbreaks.

By integrating principles from human medicine, public health, veterinary medicine, ecology, and environmental science, environmental health provides a framework for addressing these pressing global concerns. Through proactive strategies such as habitat conservation, pollution control, sustainable resource management, and climate resilience planning, this field fosters a healthier coexistence between human populations and the natural world. Understanding and prioritizing environmental health is essential for building resilient communities, preventing future pandemics, and ensuring the long-term sustainability of life on Earth.

4.1.2 The Interconnectedness of Human, Animal, and Environmental Health.

Conservation Medicine integrates ecosystem, veterinary, and public health sciences to address the complex interactions between humans, animals, and their environments. *Environmental Health* plays a key role in this framework, as it acknowledges that the degradation of ecosystems—through pollution, habitat loss, and climate change—has far-reaching consequences for both biodiversity conservation and human well-being (Daszak et al., 2001). Many emerging infectious diseases (EIDs) are zoonotic, meaning they are transmitted from animals to humans, and environmental factors often play a significant role in disease spillover. For example, deforestation, urbanization, and agricultural expansion increase human-wildlife interactions, facilitating the transmission of diseases like Ebola, HIV, and zoonotic influenza (Jones et al., 2008). By addressing environmental health, conservation medicine can help mitigate these risks, promoting not only the health of wildlife but also of humans and ecosystems.

4.1.3 Habitat Degradation, Environmental Health, and the Emergence of Zoonotic Diseases.

One of the most urgent environmental health challenges in conservation medicine is habitat degradation, a phenomenon driven by expanding human activities that disrupt natural ecosystems. As deforestation, urban sprawl, and agricultural expansion accelerate, wildlife populations experience increased stress, leading to changes in behavior, migration patterns, and social structures. These disruptions not only threaten biodiversity conservation but also heighten the risk of disease emergence and transmission.

Habitat fragmentation forces wildlife into smaller, isolated areas and increases their interactions with human populations and domestic animals. This heightened proximity creates ideal conditions for the spillover of zoonotic diseases—pathogens that can jump from animals to humans. Notable examples include Lyme disease, malaria, and West Nile virus, which have been linked to ecosystem changes that alter the distribution and abundance of host species (Plowright et al., 2017). In many cases, ecosystem imbalances, such as declines in predator populations or shifts in vector species, exacerbate the spread of infectious diseases.

Conservation Medicine highlights the critical role of habitat preservation and ecosystem restoration in mitigating these risks. Protecting intact ecosystems helps maintain natural disease regulation processes, reduces human-wildlife conflict, and preserves biodiversity conservation, which in turn supports overall environmental resilience. Strategies such as reforestation, sustainable land use planning, and wildlife corridors can mitigate the negative impacts of habitat fragmentation while promoting healthier ecosystems. By integrating conservation efforts with public health initiatives, environmental health professionals can create sustainable solutions that benefit both human and animal populations.

4.1.4 The Role of Pollution in Environmental Health and Conservation Medicine.

Pollution is a major environmental health concern with profound implications for conservation medicine, as it degrades ecosystems, disrupts biodiversity conservation, and poses severe health risks to both wildlife and human populations. Contaminants in air, water, and soil—such as industrial chemicals, agricultural runoff, and plastic waste—alter ecosystem balance and contribute to the spread of diseases. The accumulation of these pollutants in the environment threatens species survival and disrupts ecosystem services that are crucial for maintaining *Planetary Health*.

Chemical pollutants, including pesticides, heavy metals, and plastics, can bioaccumulate in the food chain, leading to severe consequences such as reproductive failures, immune system suppression, developmental abnormalities, and increased disease susceptibility in wildlife (Borg et al., 2020). Persistent organic pollutants (POPs) and endocrine-disrupting chemicals (EDCs) further compound these issues by interfering with hormonal regulation in animals, resulting in population declines and ecosystem imbalances. These toxins not only affect individual species but also have cascading effects on entire ecosystems, altering predator-prey relationships and reducing genetic diversity.

Pollution's impact extends beyond wildlife, posing serious health risks to human populations. Contaminants in drinking water, food, and air contribute to chronic illnesses such as respiratory diseases, neurological disorders, and cancers. Microplastics, for example, have been detected in human tissues, raising concerns about their long-term health effects. Additionally, air pollution—particularly fine particulate matter (PM2.5) from industrial emissions—has been linked to cardiovascular and respiratory diseases, disproportionately affecting vulnerable populations (Shaw et al., 2018).

In *Conservation Medicine*, tackling pollution is essential for safeguarding both ecosystem and human health. Strategies such as pollution control regulations, habitat remediation, sustainable agricultural practices, and plastic waste reduction can mitigate environmental contamination. Further, interdisciplinary collaboration between conservationists, public health professionals, and policymakers is vital to developing holistic solutions that prevent pollution-related health crises. By addressing pollution at its source and restoring degraded ecosystems, conservation medicine plays a crucial role in protecting biodiversity and ensuring the long-term sustainability of life on Earth.

4.1.5 Anthropocene Climate Change and Its Impact on Environmental Health.

Anthropocene, or human-driven, climate change is rapidly becoming one of the most formidable global challenges, posing grave threats to environmental health, biodiversity conservation, and the stability of ecosystems. The effects of rising global temperatures, altered precipitation patterns, and the increasing frequency and severity of extreme weather events extend far beyond environmental degradation; they profoundly influence the survival and distribution of various species, including vectors responsible for transmitting infectious diseases.

One of the most concerning impacts of climate change is its role in expanding the geographical range of disease vectors such as mosquitoes and ticks. These organisms are carriers of infectious diseases including malaria, dengue fever, and Lyme disease, which are now emerging in regions where such illnesses were previously rare or nonexistent. This expansion is fueled by warming temperatures, which create hospitable conditions for these vectors in new areas, and shifting climate dynamics that disrupt historical patterns of disease prevalence (Ebi et al., 2017).

Furthermore, the cascading effects of climate change can destabilize wildlife populations. Habitat destruction, changing resource availability, and shifting climate conditions force many species to migrate or face displacement, leading to an increased risk of zoonotic disease transmission. As humans and wildlife come into closer contact—often due to habitat encroachment, deforestation, or climate-induced migration—the likelihood of diseases being transmitted from animals to humans escalates, potentially sparking pandemics. These disruptions also have ripple effects on ecosystems, threatening their balance and integrity.

In the field of *Conservation Medicine*, there is a growing recognition of the need to address these interconnected challenges by developing and implementing robust mitigation and adaptation strategies. Conservationists and researchers advocate for measures such as habitat preservation, the establishment of wildlife corridors, and the promotion of sustainable land-use practices. These approaches aim to safeguard biodiversity conservation, reduce human-wildlife conflict, and mitigate the health risks associated with climate-induced disruptions.

In addition, combating man-made climate change requires a concerted global effort to reduce GHG emissions, invest in renewable energy, and enhance ecosystem resilience. By integrating climate-conscious policies into public health frameworks and biodiversity conservation initiatives, humanity can work to mitigate the devastating impacts of climate change on both environmental health and the intricate web of life that sustains biodiversity. This multifaceted approach is critical to ensuring a sustainable and equitable future for all species.

4.1.6 Conclusion.

Environmental Health lies at the heart of *Conservation Medicine*, as it delves into the intricate, interdependent relationships that connect humans, animals, and ecosystems. This holistic discipline acknowledges that the well-being of all species is tightly intertwined with the health of the environments they inhabit. By addressing critical factors such as habitat degradation, pollution, and the escalating impacts of climate change, *Conservation Medicine* provides a vital framework for understanding and managing the complex interplay between environmental changes and the health of both wildlife and humans.

The degradation of habitats due to deforestation, urbanization, and agricultural expansion disrupts ecosystems, displaces species, and increases human-wildlife interactions, thereby elevating the risk of zoonotic disease transmission. Pollution—whether in the form of plastics, chemical waste, or air and water contaminants—further threatens the health of ecosystems and diminishes the quality of life for countless species, including humans. Climate change amplifies these challenges by triggering extreme weather events, altering ecosystems, and destabilizing food and water resources, ultimately jeopardizing the resilience of natural and human systems.

Conservation Medicine emphasizes the importance of preventing and managing emerging infectious diseases (EIDs), many of which are linked to environmental disruptions. By identifying the connections between environmental changes and disease dynamics, this field contributes to safeguarding biodiversity conservation, protecting public health, and strengthening the resilience of ecosystems. For instance, understanding how habitat destruction facilitates the spillover of diseases from wildlife to humans can inform strategies to reduce future risks. Similarly, identifying pollution sources and their impact on ecosystems can guide remediation efforts and promote sustainable practices.

Addressing the challenges of *Environmental Health* is essential to securing a sustainable future for all species. Effective conservation strategies must prioritize the preservation and restoration of habitats, the reduction of pollution, and the mitigation of man-made climate change impacts. Initiatives such as establishing protected areas, creating wildlife corridors, and adopting sustainable land-use practices are critical to fostering biodiversity conservation and reducing human-wildlife conflicts. In addition, integrating environmental health considerations into public health policies and global development plans will further amplify efforts to address interconnected challenges.

To mitigate the far-reaching consequences of man-made climate change, global collaboration is needed to implement nature-based solutions, reduce GHG emissions, and enhance ecosystem resilience. By weaving environmental health into the fabric of conservation efforts, humanity can take meaningful steps toward reducing zoonotic disease risks, preserving biodiversity, and promoting a balanced coexistence between humans and nature. Such initiatives are pivotal in ensuring the health and sustainability of our planet for generations to come.

4.2 Environmental Health and Its Three Foundational Pillars.

4.2.1 Introduction.

As previously discussed, *Environmental Health*, plays an essential role in maintaining human health, wellness, well-being, and resiliency by examining how environmental factors, such as pollution, man-made climate change, habitat loss, and resource depletion, influence health outcomes. With growing recognition that human health is intricately linked to the health of ecosystems, the concept of *Environmental Health* encompasses a multidisciplinary approach, integrating ecology, human and veterinary medicine, public health science, and environmental sciences. It aims to minimize environmental hazards, prevent diseases, and promote healthy living conditions by fostering sustainable interactions between humans and their surroundings. As we face increasing global environmental challenges, *Environmental*

148

Health has become a central component of public health strategies and conservation efforts.

Environmental Health is built on three primary pillars: communicable and emerging infectious diseases, emergency preparedness, resilience, and response, and environmental public health principles and practices. These pillars are interconnected and form a framework for understanding how environmental factors affect human health and how to mitigate those effects.

4.2.2 Pillar 1: Communicable and Emerging Infectious Diseases.

Communicable diseases are diseases that spread from one individual to another, whether through direct contact, respiratory droplets, contaminated food or water, or vectors such as mosquitoes or ticks. These diseases, such as influenza, tuberculosis, and HIV/AIDS, have long been central to global public health concerns. However, emerging infectious diseases (EIDs)—newly identified diseases or those re-emerging after a period of absence—pose an even greater challenge. Environmental factors, such as deforestation, man-made climate change, and urbanization, are driving the spread of EIDs and exacerbating the risk of zoonotic diseases, which are transmitted from animals to humans.

As human activities encroach on wildlife habitats, we increase the likelihood of human-wildlife interactions, providing opportunities for pathogens to spill over into human populations. Zoonotic diseases, including Ebola, SARS, and more recently, COVID-19, illustrate how environmental changes can create conditions that lead to the emergence of diseases that affect both humans and animals. Deforestation, land-use changes, and wildlife trafficking contribute to these changes, disrupting ecosystems and altering pathogen dynamics (Daszak et al., 2001).

Man-made climate change plays a significant role in modifying the distribution of disease vectors and pathogens. For example, rising global temperatures and changing precipitation patterns are expanding the range of mosquitoes that transmit diseases such as malaria, dengue, and Zika virus (Gage et al., 2008). Warmer climates allow these vectors to thrive in areas previously unsuitable for their survival. Furthermore, shifting ecosystems create opportunities for previously isolated populations of animals, insects, and pathogens to come into contact, further increasing the risk of cross-species transmission.

Addressing communicable and emerging infectious diseases within *Environmental Health* requires a multidisciplinary approach. Effective disease surveillance and early response, monitoring of animal and human health, vaccination campaigns, and habitat preservation are all vital components. Strengthening public health and environmental policies that regulate wildlife trade, managing ecosystems to reduce the risk of pathogen spillover, and improving sanitation and hygiene are also critical measures to mitigate the risks posed by these diseases (Jones et al., 2008).

4.2.3 Pillar 2: Emergency Preparedness, Resilience, and Response.

Emergency preparedness refers to the activities, policies, and plans designed to anticipate and minimize the health impacts of environmental hazards and disasters. As man-made climate change accelerates, the frequency and intensity of natural disasters—such as floods, hurricanes, wildfires, and droughts—are increasing, resulting in significant public health consequences. *Environmental Health* practitioners must work proactively to assess risk factors, identify vulnerable and marginalized populations, and develop strategies for effective response and recovery.

Resilience is the ability of a community or system to absorb shocks and recover from disasters. It involves strengthening infrastructure, ensuring healthcare delivery systems are equipped to handle crises, and building community capacity to respond to and recover from *Environmental Health* challenges. For example, cities that invest in climate-resilient infrastructure, such as green roofs to reduce heat, or flood defenses to prevent waterborne diseases, can better withstand extreme weather events. Resilience also extends to environmental systems—restoring natural ecosystems, such as wetlands, can mitigate the impact of flooding and improve water quality, ultimately protecting the public's health.

Response involves the immediate actions taken during and after a disaster to mitigate health impacts. These actions include providing holistic, integrated, and comprehensive medical services, ensuring access to clean water and sanitation, conducting disease surveillance, and addressing mental health needs. The response to *Environmental Health* emergencies also includes the coordination of relief efforts, appropriate and equitable resource allocation, reinforcing supply chain activity, and ensuring the safety and well-being of affected populations.

Integrating *Environmental Health* into emergency preparedness and response efforts is essential to reducing the health risks associated with disasters. The *Sendai Framework for Disaster Risk Reduction,* adopted by the United Nations in 2015, emphasizes the need to reduce disaster risks by strengthening essential public health functions and systems, improving environmental services management, and ensuring that communities are equipped to handle crises (UNDRR, 2015). A holistic approach that incorporates *Environmental Health*, climate mitigation and adaptation strategies, and community resilience is crucial to managing the increasing environmental risks posed by climate change and natural disasters.

4.2.4 Pillar 3: Environmental Public Health Principles and Practices.

Environmental public health principles and practices form the backbone of the field of *Environmental Health*. These guiding principles aim to reduce environmental hazards and foster healthy living spaces through proactive measures, regulatory frameworks, and active community participation. Core areas of focus in environmental public health services include ensuring clean air and water quality, effective waste management, safe handling of hazardous materials, and promoting environmental justice to address disparities and ensure equitable access to healthy environments.

Air quality is one of the most critical environmental health concerns. Poor air quality is a leading cause of respiratory and cardiovascular diseases, such as asthma, chronic obstructive pulmonary disease (COPD), and heart disease. Industrial emissions, transportation, and the burning of fossil fuels contribute to air pollution, particularly in urban areas. According to the World Health Organization (WHO), air pollution is responsible for millions of premature deaths each year (WHO, 2021). Addressing air quality requires public policies that regulate emissions, promote clean energy, and invest in green infrastructure to reduce pollutants.

Water quality is another vital aspect of *Environmental Health*. Contaminated water can lead to a wide range of diseases, including cholera, dysentery, and typhoid fever. Unsafe drinking water, inadequate sanitation, and improper waste disposal contribute to waterborne diseases, particularly in low-income regions. The World Health Organization estimates that more than two billion people lack access to safe drinking water, posing a significant public health challenge (WHO, 2017). Ensuring access to clean water requires investments in water infrastructure, sanitation systems, hygienic practices, and pollution control measures.

Waste management is crucial for preventing environmental contamination and protecting human health. Improper disposal of waste, including plastics, chemicals, and hazardous substances, can lead to soil, water, and air pollution, contributing to the spread of diseases and environmental degradation. Effective waste management practices, such as recycling, composting, and hazardous waste disposal, are essential for reducing health risks.

Environmental justice ensures that all communities, regardless of socio-economic status, have equitable access to a healthy environment. Historically, marginalized and vulnerable populations have been disproportionately exposed to environmental hazards, such as industrial pollution, poor air and water quality, and inadequate sanitation. *Environmental Health* policies must prioritize equitable access to clean and safe living conditions for everyone around-the world (Kusnick et al., 2017).

4.2.5 Conclusion.

Environmental Health plays a pivotal role in both *Conservation Medicine* and public health by emphasizing the critical connections between human well-being, ecosystem health, and environmental factors. As we continue to face the mounting pressures of man-made climate change, environmental degradation, and biodiversity loss, the need to understand and address the three (3) foundational pillars of *Environmental Health* becomes increasingly urgent. These pillars—communicable and emerging infectious diseases, emergency preparedness, resilience and response, and environmental public health principles and practices—serve as a framework for protecting human health and ensuring the sustainable management of ecosystems.

Communicable and emerging infectious diseases are major public health challenges that arise when environmental disruptions, such as deforestation, urbanization, and climate change, create conditions favorable for the spread of pathogens. By fostering a deeper understanding of how environmental factors influence the emergence and transmission of diseases, societies can better mitigate the risks of zoonotic outbreaks, such as Ebola, Zika, and COVID-19. Protecting ecosystems, regulating wildlife trade, and implementing disease surveillance systems are key strategies in reducing the spread of infectious diseases, especially in a world where pathogens can easily cross borders due to globalization.

Emergency preparedness, resilience, and response are crucial in addressing the growing frequency and intensity of environmental disasters. Whether caused by natural phenomena like floods, wildfires, or hurricanes, or driven by human activities such as pollution and industrial accidents, these crises have significant impacts on public health. Strengthening emergency response mechanisms, building resilience through infrastructure improvements, and promoting community engagement in disaster preparedness are essential steps in minimizing harm during environmental emergencies. Equally important is ensuring that healthcare delivery systems are equipped to respond equitably, efficiently, and effectively to these challenges, with a focus on reducing health disparities that disproportionately affect marginalized and vulnerable populations.

Environmental public health principles and practices, which include the management of air and water quality, waste disposal, and hazardous materials, are vital for safeguarding both human and eco-system health. Public health and *Environmental Health* policies that regulate pollutants, promote clean energy, and address environmental justice help reduce exposure to harmful substances that contribute to chronic diseases such as asthma, cardiovascular disease, and cancer. In addition, the integration of sustainable land-use practices and biodiversity conservation into public health strategies and practices helps ensure that natural ecosystems continue to provide essential services, such as water filtration, air purification, and climate regulation.

By adopting an approach that combines prevention, promotion, protection, surveillance, regulation, and sustainable preparedness practices, *Environmental Health* interventions can help address the root causes of *Environmental Health* risks. Proactive measures, such as reducing GHG emissions, conserving natural habitats, and implementing sustainable agricultural practices, are essential for protecting both human populations and biodiversity conservation. Furthermore, by promoting the use of empirically-derived evidence-based public health strategies and services, public health policies can be created that not only protect individuals from environmental hazards but also foster the restoration and preservation of ecosystems, ultimately benefiting all living organisms.

The integration of *Environmental Health* into global health policies and disaster response efforts will be instrumental in building capacity and resilience to the impacts of man-made climate change and environmental degradation. Addressing health-related social needs and other determinants of health, such as poverty, access to clean water, and safe housing, will ensure that all communities have the resources to adapt to changing environmental conditions. Collaborative efforts among governments, international organizations, healthcare providers, and local communities are essential for developing solutions that not only safeguard public health but also preserve the ecosystems on which we all depend now and for future generations to come. As the global public-at-large moves toward a more sustainable future, the integration of *Environmental Health* into broader public health and *Environmental Health* policies and practices will be critical in ensuring that both human and ecosystem health are prioritized in the face of global environmental challenges.

The intersection of *Environmental Health, Conservation Medicine,* and public health highlights the importance of addressing environmental factors as critical drivers to improving the health, wellness, well-being, and resiliency of individuals and populations. Through holistic, comprehensive and integrated approaches, societies world-wide can reduce the risks associated with man-made climate change, habitat loss, and pollution, while promoting long-term sustainability for both human and planetary health. Prioritizing *Environmental Health* within global public health and *Environmental Health* policies will be essential for creating a healthier, more resilient future for all species on Earth.

4.3 Case Study: The Exposome, the Amazon Rainforest, Environmental Health and Conservation Medicine.

4.3.1 Introduction.

The concept of the exposome refers to the cumulative environmental exposures that an individual experiences throughout their lifetime, encompassing everything from air and water quality to determinants of health and lifestyle factors (Wild, 2005). Unlike the human genome, which focuses solely on an individual's genetic makeup, the exposome takes a broader view, incorporating all external and internal factors that influence individual health and wellness. In recent years, the concept of the exposome has gained traction in both human and environmental health fields, particularly in *Conservation Medicine*. *Conservation Medicine* seeks to understand the complex interactions between human health, animal health, and ecosystems, and the exposome framework provides a useful lens for understanding how environmental exposures contribute to the emergence of diseases and the loss of biodiversity conservation. This case study explores how the exposome model, within the context of the Amazon Rainforest, can enhance conservation medicine practices, focusing on an example of emerging infectious diseases (EIDs) and their implications for both human and wildlife health.

4.3.2 Environmental Health and the Exposome: An Overview.

The exposome (Wild, 2005) is a comprehensive measure of the environmental factors that impact human health across one's life course health development (Halfon et al, 2014), starting from pre-conception and continuing throughout adulthood to end-of-life. It includes environmental health factors such as:

1. *Chemical Exposures:* Pollution, pesticides, and heavy metals in the environment.
2. *Physical Exposures:* Radiation, temperature, and noise pollution.
3. *Biological Exposures:* Pathogens, allergens, and disease vectors.
4. *Social and Behavioral Factors:* Lifestyle choices, socioeconomic status, and access to healthcare.

These exposures interact with an individual's genetic makeup to shape health outcomes, making the exposome a powerful tool for understanding the effects of the environment on disease development. In the context of *Conservation Medicine*, the exposome approach emphasizes how environmental changes, such as deforestation, man-made climate change, and habitat destruction, can have cascading effects on *Environmental Health*, both human and wildlife health (Sanchirico et al., 2020).

4.3.3 The Amazon Rainforest, Environmental Health, and Conservation Medicine.

One of the most urgent challenges in *Conservation Medicine* is the rise of zoonotic diseases—illnesses transmitted from animals to humans. These diseases pose significant threats to global environmental health, particularly as environmental changes and human activities accelerate their emergence and spread. Among the regions most affected by this dynamic is the Amazon Rainforest, a biodiversity hotspot and vital component of the Earth's ecosystem balance. Often referred to as the *lungs of the Earth*, the Amazon Rainforest plays a critical role in regulating the global climate, sustaining countless species, and supporting biodiversity conservation efforts.

The Amazon Rainforest harbors an extraordinary variety of flora and fauna, many of which remain unexplored and potentially hold valuable resources for medicine and science. However, ongoing deforestation, driven by agricultural expansion, illegal logging, and infrastructure development, is transforming this invaluable ecosystem in profoundly negative ways. The destruction of forests not only reduces the Amazon's capacity to absorb carbon dioxide and mitigate climate change but also disrupts the delicate balance and health of wildlife populations.

These environmental disruptions create conditions conducive to the spillover of zoonotic diseases. Habitat destruction displaces wildlife species, forcing them into closer contact with human populations, which increases the likelihood of disease transmission. For example, pathogens carried by bats or rodents may spread more easily when these animals lose their natural habitats and interact with humans and domestic animals. Such spillover events have the potential to lead to epidemics or even pandemics, as seen in recent years.

Beyond the health risks, deforestation also disrupts the Amazon's ability to provide essential ecosystem services, such as water regulation, soil fertility, and climate stabilization. The degradation of these systems jeopardizes not only the health of wildlife but also the livelihoods and well-being of local communities that depend on the rainforest's resources for survival.

To address these challenges, *Conservation Medicine* calls for urgent and integrative *Environmental Health* solutions. Strategies such as enforcing anti-deforestation laws, promoting sustainable agricultural practices, and investing in reforestation efforts are critical to preserving the Amazon's ecological integrity. Additionally, monitoring and mitigating zoonotic disease risks require coordinated efforts between governments, scientists, and local communities. Developing early warning systems, enhancing surveillance of wildlife health, and reducing human-wildlife interactions can help minimize the chances of disease spillover.

Ultimately, safeguarding the Amazon Rainforest is not just about protecting a single region—it is about securing global environmental health, biodiversity conservation, and ecosystem services stability. The inextricable interconnectedness of humans, wildlife, and ecosystems underscores the importance of *Conservation Medicine* in addressing complex challenges and ensuring the resilience of the planet's most vital habitats.

4.3.4 The Exposome and the Amazon Rainforest.

In the case of the Amazon Rainforest, multiple components of the exposome—the totality of environmental exposures that impact human health—intersect, creating conditions ripe for the emergence and spread of zoonotic diseases. Deforestation, one of the most prominent drivers of environmental change in the Amazon, forces wildlife into closer proximity to human populations. This increased contact significantly heightens the likelihood of zoonotic pathogens crossing from animals to humans, potentially sparking outbreaks of diseases with serious public health implications.

Human activities such as illegal logging, agricultural expansion, mining, and infrastructure development in the Amazon exacerbate these risks. As forested areas are cleared, the natural habitats of many animal species shrink, leading to increased interaction between humans, wildlife, and domestic animals. These interactions create opportunities for disease vectors—organisms like mosquitoes, ticks, and rodents that transmit infectious diseases—to thrive in disturbed ecosystems. Pathogens that were once confined to the deep forests now find new routes to spread, elevating the risk of diseases such as malaria, dengue fever, and Zika virus becoming more widespread.

Adding another layer of complexity, man-made climate change intensifies these challenges by altering environmental conditions that directly affect the distribution and behavior of vectors. Rising temperatures, shifting precipitation patterns, and changes in humidity create favorable conditions for vectors like mosquitoes and ticks to expand their geographical range into previously unsuitable areas. This means diseases such as malaria, dengue, and Zika may spread to new regions, including urban centers and higher altitudes, where populations have limited immunity and healthcare infrastructure may be inadequate to address the surge in cases (Hoffmann et al., 2019).

Moreover, the combined impact of deforestation and climate change disrupts the delicate balance of the Amazon's ecosystem, compromising its ability to provide essential services like carbon storage, water regulation, and climate stabilization. These disruptions have far-reaching consequences for both human and environmental health, affecting not just local communities but also the global population that depends on the Amazon's role in regulating the planet's climate.

To mitigate these risks, integrative strategies are essential. Protecting the Amazon from further deforestation, enforcing regulations to curb illegal activities, and promoting sustainable agricultural and land-use practices are critical steps. Additionally, strengthening disease surveillance systems in the region can help identify and respond to emerging zoonotic threats early. Addressing climate change through global cooperation to reduce greenhouse gas emissions and enhance ecosystem resilience is equally vital in minimizing the cascading effects on public health and biodiversity.

The interconnected challenges in the Amazon Rainforest underscore the importance of a multidisciplinary and multisectoral approach that combines environmental science, public health, and biodiversity conservation efforts. Protecting this unique and irreplaceable ecosystem is not only a moral imperative but also a practical necessity to safeguard global health and sustain life on Earth.

4.3.5 Yellow Fever: Zoonotic Disease Transmission in the Amazon Rainforest.

The presence of yellow fever in the Amazon Rainforest exemplifies how environmental changes can dramatically influence zoonotic disease transmission. Yellow fever, a viral illness transmitted by mosquitoes, primarily affects monkeys in the rainforest ecosystem. However, when their habitats are disrupted, the virus can spill over into human populations, leading to serious outbreaks. In the Amazon, forest fragmentation caused by deforestation—driven by illegal logging, agricultural expansion, and infrastructure development—has increased the chances of humans and mosquitoes coming into contact with primates that carry the yellow fever virus. This heightened interaction underscores the role of environmental disruptions as a catalyst for zoonotic spillover events, highlighting the interconnectedness of animal and human health (Gubler, 2018).

Deforestation and climate change act as twin forces exacerbating this risk. Fragmenting and destroying natural habitats not only displace wildlife but also forces them into closer proximity to humans and other species, creating opportunities for vectors like mosquitoes to transmit diseases across species boundaries. Moreover, man-made climate change compounds this challenge by altering environmental conditions—such as temperature and precipitation patterns—making areas more suitable for disease vectors. These changes expand the habitats of mosquitoes and other vectors, spreading diseases like malaria, dengue fever, and yellow fever to new regions, many of which may be ill-prepared to handle such outbreaks.

Yellow fever is not an isolated example. Other major zoonotic diseases, such as Ebola and SARS-CoV-2 (responsible for the COVID-19 pandemic), have similarly been linked to disruptions of natural habitats. For instance, the encroachment of human settlements into forests and wildlife territories has played a significant role in creating pathways for disease transmission. The destruction of forests eliminates the natural barriers that once kept humans and wildlife separated, making spillover events more frequent and, consequently, more dangerous.

The exposome framework provides a comprehensive lens through which to understand these phenomena. This framework examines the cumulative impact of environmental exposures, such as habitat loss, deforestation, and climate change, on human health. In the cases of Ebola and SARS-CoV-2, the exposome helps elucidate how anthropogenic activities disrupt ecosystems and alter the dynamics of disease transmission, paving the way for zoonotic pathogens to emerge and spread on a global scale (Daszak et al., 2020).

Addressing these interconnected challenges requires a multidimensional and multisectoral approach. Efforts must focus on preventing further habitat destruction by enforcing anti-deforestation laws and promoting sustainable practices in land use and agriculture. Strengthening disease surveillance and response systems to monitor the health of wildlife and detect zoonotic threats early can also minimize the risk of spillover events. Additionally, combating man-made climate change through global cooperation to reduce GHG emissions is essential to stabilizing ecosystems and curbing the spread of vector-borne diseases.

Ultimately, these actions not only protect vulnerable human populations but also preserve the intricate balance of ecosystems that sustain life on Earth. Understanding the links between environmental changes and zoonotic diseases serves as a call to safeguard both the Amazon Rainforest and global health, emphasizing the importance of a harmonious coexistence between humans and the natural world.

4.3.6 Ecosystem Destruction: Wildlife Health and Loss of Biodiversity Conservation.

The loss of biodiversity conservation due to habitat destruction profoundly disrupts the delicate ecosystem relationships that serve as natural barriers against disease. In ecosystems like the Amazon Rainforest, such biodiversity conservation loss can have cascading effects on both wildlife and human health. Species such as bats and primates, for example, play critical roles in regulating the populations of disease vectors or controlling organisms that may harbor pathogens. The decline or disappearance of these species weakens these natural controls, which can result in the proliferation of other species—such as rodents or insects—that are more likely to carry and transmit pathogens. This imbalance increases the risk of zoonotic diseases spilling over into human populations, posing a significant threat to global health.

The loss of biodiversity conservation is particularly alarming for ecosystem health as a whole. Biodiverse ecosystems are more resilient, meaning they are better able to withstand and recover from disturbances such as disease outbreaks, natural disasters, or environmental changes. When biodiversity conservation is diminished, ecosystems lose their ability to function effectively, leading to further degradation and diminished capacity to support life, including human communities that depend on these ecosystems for essential services like clean water, air, and food.

The exposome framework offers a comprehensive way to understand these interconnected risks. It highlights how environmental exposures—ranging from chemical pollutants and infectious pathogens to social and behavioral factors—cumulatively shape environmentally-focused health outcomes for both wildlife and humans.

For instance, human activities such as poaching, illegal logging, and land conversion for agriculture directly increase the likelihood of human-wildlife interactions. These closer interactions break down the natural barriers that previously minimized the risk of disease transmission, enabling pathogens to jump between species more easily. This process has been implicated in numerous outbreaks of zoonotic diseases (Burgess et al., 2017).

Moreover, habitat destruction and human encroachment into previously undisturbed areas create "hot spots" of disease emergence. Forest fragments and disturbed ecosystems often concentrate both wildlife and vectors like mosquitoes, further compounding the risk of spillover events. Anthropocene climate change exacerbates these challenges by altering the distribution and behavior of disease vectors, spreading diseases into new areas where populations may lack immunity or infrastructure to cope with outbreaks.

To address these interconnected challenges, preserving biodiversity conservation must become a cornerstone of public health and environmental policy. Strategies such as protecting and restoring habitats, establishing wildlife corridors, and enforcing laws against illegal logging and poaching are essential to safeguarding biodiversity. In tandem, reducing GHG emissions to mitigate climate change will help stabilize ecosystems and limit the spread of vector-borne diseases.

Additionally, fostering community engagement and environmental health education is crucial to reducing harmful activities that threaten biodiversity conservation. Integrated disease surveillance systems that monitor both wildlife and human health can provide early warnings and prevent outbreaks before they escalate. By adopting a holistic and integrated approach that recognizes the links between biodiversity conservation, ecosystem health, and public health, we can address the root causes of these global challenges and work toward a healthier, more sustainable future for all life on Earth.

4.3.7 Lessons Learned: Integrating the Exposome into Environmental Health and Conservation Medicine.

Conservation Medicine provides a unique framework for addressing the environmental health challenges posed by environmental degradation, as it recognizes the inextricable interconnectedness of human, animal, and environmental health. The exposome model enriches this approach by offering a comprehensive understanding of how environmental factors influence environmental health outcomes. By integrating exposome data into conservation medicine practices, researchers and public health officials can develop more effective strategies and actions to prevent zoonotic disease spillover and mitigate the effects of environmental degradation.

In practice, this means:

1. *Environmental monitoring:* Continuously tracking environmental changes and their impacts on disease dynamics, as well as assessing the health of ecosystems.
2. *Ecosystem restoration:* Implementing efforts to restore natural habitats to reduce human-wildlife conflicts and preserve biodiversity conservation.
3. *Disease surveillance and rapid response:* Strengthening surveillance systems for both wildlife and human populations to detect emerging diseases early and respond quickly, equitably, and effectively.
4. *Sustainable land use:* Promoting sustainable land-use practices that minimize human impact on natural habitats and reduce the likelihood of zoonotic disease transmission.

4.3.8 Conclusion.

The integration of the exposome model into *Environmental Health* and *Conservation Medicine* enhances our understanding of how environmental exposures contribute to emerging infectious diseases (EIDs), wildlife health, and biodiversity loss. In the case of the

Amazon Rainforest, deforestation and man-made climate change not only impact the ecosystem but also create pathways for diseases to spill over from animals to humans, resulting in public health and environmental health crises. By adopting an exposome-driven approach to *Environmental Health* and *Conservation Medicine*, societies can develop more effective strategies for preventing disease transmission, protecting biodiversity conservation, and ensuring the health of both human and animal populations in a rapidly changing world.

4.4 References.

1. Alonso, P., et al. (2011). Deforestation and its impact on malaria transmission in the Amazon basin of Brazil. *Malaria Journal*, 10(1), 178.
2. Borg, D., et al. (2020). The effects of environmental pollutants on wildlife health. *Environmental Toxicology and Chemistry*, 39(6), 1483-1493.
3. Borg, D., Lundh, T., Lindh, C. H., & Sundström, B. (2020). Environmental pollutants and their effects on wildlife: An overview of exposure and health risks. *Environmental Research*, 182, 109010.
4. Burgess, E. A., et al. (2017). Conservation Medicine: a novel approach to ecosystem health. *Environmental Health Perspectives*, 125(6), 645-652.
5. Centers for Disease Control and Prevention (CDC). (2023). *Environmental Health*.
6. Daszak, P., et al. (2020). Emerging infectious diseases and the risk to biodiversity: A global framework for surveillance. *Science*, 358(6359), 604-609.
7. Daszak, P., et al. (2001). Emerging infectious diseases of wildlife—threats to biodiversity and human health. *Science*, 287(5452), 443-449.
8. Ebi, K. L., et al. (2017). The impact of climate change on human health. *The Lancet*, 389(10079), 619-631.

9. Gage, K. L., et al. (2008). Climate and infectious disease: climate effects on vector-borne diseases. *The Lancet Infectious Diseases*, 8(12), 713-723.

10. Gubler, D. J. (2018). The mosquito-borne diseases of the Amazon. *The Lancet Infectious Diseases*, 18(6), 627-634.

11. Halfon, N., Larson, K., Lu, M., Tullis, E., & Russ, S. (2014). Lifecourse health development: past, present and future. *Maternal and child health journal, 18*(2), 344–365.

12. Hoffmann, B., et al. (2019). Impacts of climate change on vector-borne diseases in the Amazon. *Global Health Action*, 12(1), 1541806.

13. Jones, K. E., et al. (2008). Global trends in emerging infectious diseases. *Nature*, 451(7181), 990-993.

14. Karesh, W. B., et al. (2012). Ecology of zoonoses: natural and unnatural histories. *The Lancet*, 380(9857), 1936-1945.

15. Kusnick, J. M., et al. (2017). Environmental health disparities and environmental justice. *American Journal of Public Health*, 107(9), 1459-1460.

16. National Institute of Environmental Health Sciences (NIEHS). (2023). *What is Environmental Health?*

17. Plowright, R. K., et al. (2017). Pathways to zoonotic spillover. *Nature Reviews Microbiology*, 15(8), 502-510.

18. Sanchirico, J. N., et al. (2020). Integrating the exposome into conservation medicine. *Conservation Science and Practice*, 2(10), e267.

19. Shaw, J. R., et al. (2018). Impact of environmental pollutants on wildlife health. *Environmental Pollution*, 235, 282-293.

20. Shaw, S. D., Berger, M. L., Brenner, D., et al. (2018). *Persistent organic pollutants and endocrine-disrupting chemicals in wildlife and humans: Trends, toxicological effects, and challenges.* Environmental Science & Technology, 52(6), 3310-3323.

21. United Nations Office for Disaster Risk Reduction (UNDRR) (2015). Sendai Framework for Disaster Risk Reduction 2015–2030. United Nations Office for Disaster Risk Reduction.

22. Wild, C. P. (2005). The exposome: from concept to utility. *International Journal of Epidemiology*, 34(5), 1119-1127.
23. World Health Organization (WHO). (2017). Water Quality and Health. *World Health Organization*.
24. World Health Organization (WHO). (2018). Environmental health in emergencies. *World Health Organization*.
25. World Health Organization (WHO). (2021). Air quality and health. *World Health Organization*.
26. World Health Organization (WHO). (2023). *Environmental Health*.

5.0

Conservation Medicine in Action.

Conservation Medicine is a dynamic and interdisciplinary field that bridges the gap between animal, human, and ecosystem health. By recognizing the inextricable interconnectedness of all living systems, it seeks innovative solutions to address the challenges posed by habitat loss, emerging infectious diseases, and Anthropocene climate change. *Conservation Medicine in Action* highlights the practical applications of this approach, showcasing efforts to protect biodiversity conservation, restore ecosystems, and promote sustainable coexistence between humans and wildlife. It's an exploration of how real science and empathy unite to safeguard the planet's health for future generations.

5.1 Introduction.

As previously described, *Conservation Medicine* is an interdisciplinary field that examines the complex relationships between human health, animal health, and ecosystem health. Rooted in the *One Health* approach, *Conservation Medicine* integrates knowledge from human and veterinary medicine, public health, ecology, and environmental science to address emerging infectious diseases, biodiversity loss, and environmental degradation (Aguirre et al., 2002). As man-made climate change, habitat destruction, and globalization accelerate, *Conservation Medicine* plays a crucial role in mitigating zoonotic dis-

168

ease transmission, promoting ecosystem resilience, and safeguarding both human and wildlife populations (Daszak et al., 2000).

A defining aspect of *Conservation Medicine* is its focus on the intricate web of ecosystem interactions that influence health outcomes. For example, habitat destruction and deforestation have been linked to increased human-wildlife contact, facilitating the spillover of pathogens such as Ebola, Nipah virus, and coronaviruses (Jones et al., 2008). Similarly, man-made climate change alters disease dynamics, influencing the geographic distribution of vector-borne diseases like malaria, Lyme disease, and dengue fever (Patz et al., 2005). *Conservation Medicine* provides a framework for understanding these existential threats to all living things and implementing strategies to mitigate their impact through disease surveillance and response, habitat preservation, and policy interventions.

Successful conservation medicine initiatives often involve multisectoral collaboration among scientists, policymakers, and local communities. One example is the *PREDICT project*, which has helped identify high-risk areas for zoonotic spillover by monitoring wildlife and human disease interactions (Carroll et al., 2018). It was truly an ambitious initiative under *USAID's Emerging Pandemic Threats program*, launched in 2009 and led by the UC Davis *One Health* Institute. The project aimed to strengthen global capacity for detecting and responding to viruses with pandemic potential, particularly those that could spill over from animals to humans. Over its decade-long operation, PREDICT identified nearly 1,000 new viruses, including novel strains of Ebola and coronaviruses, and trained thousands of professionals worldwide in One Health surveillance. It also enhanced laboratory systems in over 30 countries, focusing on regions with high biodiversity and significant human-animal interaction. The project was instrumental in advancing global health security and pandemic preparedness before its closure in 2020.

Additionally, *Conservation Medicine* contributes to sustainable development by promoting practices that protect biodiversity conservation while ensuring food security and the public's health, such as integrating livestock management with wildlife conservation to reduce disease transmission (Kock, 2014).

As global health challenges become increasingly complex, *Conservation Medicine* offers a proactive, holistic, integrated, and comprehensive approach to addressing them. By recognizing the interdependence of human, animal, and ecosystem health, this multisectoral field provides critical insights and practical solutions to enhance global health security and ecosystem stability. Moving forward, investment in conservation medicine research, education, practice, and policy implementation will be essential to mitigating the health risks associated with environmental change and loss of biodiversity conservation.

5.2 Wildlife Disease Surveillance, Monitoring, and Rapid Response.

5.2.1 Introduction.

Wildlife disease surveillance, monitoring, and rapid response are cornerstone practices within the field of *Conservation Medicine*. These activities serve as critical early warning systems for identifying and mitigating emerging infectious diseases (EIDs) that pose significant threats to biodiversity conservation, livestock health, and human populations. Through systematic observation and analysis of disease incidence, prevalence, and transmission dynamics within wildlife populations, conservation medicine practitioners are equipped to uncover complex interactions between pathogens, host species, and environmental factors.

By examining ecosystem drivers such as habitat loss, climate change, and human-wildlife interactions, practitioners can pinpoint conditions that foster disease emergence and spread. This proactive approach enables the identification of high-risk areas and populations, allowing for the development of targeted public health interventions and conservation strategies. Furthermore, such efforts enhance our understanding of zoonotic disease risks—instances where pathogens cross species barriers from animals to humans—thereby strengthening global health security frameworks.

The integration of wildlife disease surveillance into broader conservation and public health initiatives exemplifies the holistic *One Health* approach, which acknowledges the interdependence of human, animal, and environmental health. By addressing these connections, *Conservation Medicine* contributes to safeguarding ecological balance while simultaneously protecting the well-being of diverse species, including our own (Grooten & Almond, 2018).

5.2.2 Wildlife Disease Surveillance, Monitoring, and Rapid Response: Why is it Important?

Wildlife disease surveillance serves as a critical tool in the early detection and prevention of zoonotic diseases, enabling interventions before they escalate into widespread outbreaks. With the emergence of high-impact pathogens such as the Ebola virus, highly pathogenic avian influenza, and various coronaviruses—including SARS-CoV-2—there is a growing urgency to prioritize the continuous monitoring of wildlife health. This need is especially pressing in regions where human-wildlife interactions are intensifying due to factors such as habitat encroachment, agricultural expansion, and climate change, which disrupt ecosystems and create conditions conducive to spillover events (Daszak et al., 2000).

Comprehensive surveillance programs are essential for collecting and analyzing critical data on disease dynamics. Such programs allow researchers and public health officials to identify emerging patterns, assess potential risks of cross-species transmission, and implement targeted control measures to mitigate these threats (Carroll et al., 2018). Moreover, by integrating advanced technologies such as genomic sequencing and geographic information systems (GIS), wildlife surveillance can enhance predictive modeling, enabling authorities to forecast potential hotspots for zoonotic disease emergence. Proactive strategies, rooted in robust surveillance systems, not only protect the public's health but also contribute to preserving bio-diversity conservation and maintaining ecological balance.

5.2.3 Methods of Disease Surveillance and Monitoring.

Effective wildlife disease surveillance employs various methodologies, including:

1. *Passive Surveillance* – This involves the reporting of disease cases in wildlife by researchers, veterinarians, and local communities. Although cost-effective, passive surveillance often underestimates disease prevalence due to underreporting and lack of systematic data collection (Kuiken et al., 2005).
2. *Active Surveillance* – Involves targeted sampling of wildlife populations to detect asymptomatic carriers and identify emerging pathogens. Techniques include biological sampling (e.g., blood, feces, and nasal swabs), necropsies, and genetic sequencing of pathogens (Anthony et al., 2013).
3. *Remote Sensing and Technology Integration* – Advances in satellite imaging, environmental DNA (eDNA) analysis, and AI-driven diagnostics have enhanced wildlife disease surveillance. For example, remote sensing can track environmental changes linked to disease outbreaks, such as deforestation and climate anomalies that influence vector-borne diseases (Gilbert et al., 2014).
4. *Community-Based Monitoring* – Engaging local communities in disease surveillance helps bridge knowledge gaps, as indigenous and rural populations often have firsthand observations of wildlife health changes. Programs such as *EcoHealth Alliance's PREDICT* initiative integrate community knowledge with scientific monitoring to enhance zoonotic disease preparedness (Mazet et al., 2016).

5.2.4 Case Studies in Wildlife Disease Monitoring.

1. *PREDICT Project:* This global initiative, led by the United States Agency for International Development (USAID), monitored wildlife diseases in hotspots for zoonotic spillover. The project identified over 1,000 novel viruses, including coronaviruses related to SARS-CoV-2, underscoring the importance of early surveillance in pandemic prevention (Carroll et al., 2018).

2. *White-Nose Syndrome in Bats:* Continuous monitoring of North American bat populations has helped track the spread of *Pseudogymnoascus destructans*, a fungus causing White-Nose Syndrome. Surveillance data informed conservation efforts, such as habitat protection and artificial hibernation sites, to mitigate bat mortality (Blehert et al., 2009).

3. *Avian Influenza Surveillance*: Ongoing surveillance of migratory birds has provided critical insights into the spread of highly pathogenic avian influenza (HPAI), enabling early warnings and biosecurity measures to prevent outbreaks in domestic poultry and human populations (Gilbert et al., 2014).

5.2.5 Challenges and Future Directions of Wildlife Disease Surveillance.

Although wildlife disease surveillance is vital for preventing zoonotic outbreaks and maintaining ecosystem balance, it encounters numerous challenges that hinder its effectiveness. Limited funding remains one of the most critical barriers, as sustained financial support is necessary to develop and implement comprehensive disease surveillance systems. Logistical constraints, particularly in remote and biodiverse regions, further complicate efforts to monitor wildlife health. Accessing these areas requires significant resources and coordination, which is often difficult to achieve. Additionally, there is a pressing need for greater international collaboration to address cross-border disease threats and harmonize surveillance efforts (Karesh et al., 2012).

To overcome these challenges, future initiatives should focus on several key priorities:

1. *Expanding Global Disease Surveillance Networks with Real-Time Data Sharing*:
    Strengthening global networks will facilitate the rapid exchange of critical information, enabling timely responses to emerging threats. This requires interoperable data systems that allow real-time sharing of pathogen and disease-related findings across countries and organizations.

2. *Enhancing Technological Capabilities for Rapid Pathogen Detection*:
    Innovations such as portable diagnostic tools, genomic sequencing, and artificial intelligence-powered analytics can significantly improve the speed and accuracy of pathogen identification. Leveraging technology ensures that surveillance efforts keep pace with the evolving nature of zoonotic diseases.

3. *Strengthening Interdisciplinary and Multisectoral Collaboration*:
    The integration of expertise from diverse fields—including ecology, human and veterinary medicine, epidemiology, and public health—is crucial for a holistic and integrated approach to wildlife disease surveillance. Collaborative frameworks that encourage shared insights and coordinated strategies will lead to more effective interventions.

By addressing these priorities, the global community can build a more resilient wildlife disease surveillance, monitoring, and rapid response system, ultimately reducing the risk of zoonotic outbreaks and safeguarding both human health and biodiversity.

5.2.6 Conclusion.

Wildlife disease surveillance, monitoring, and rapid response system serves as a foundational pillar of *Conservation Medicine*, offering critical insights into the intricate and dynamic interactions between human, animal, and environmental health. By systematically tracking disease patterns in wildlife populations, researchers can identify potential zoonotic threats, understand ecosystem drivers of disease emergence, and develop targeted strategies to mitigate outbreaks before they escalate into global health crises. Strengthening disease surveillance systems through enhanced data collection, real-time reporting, and international collaboration is essential for improving early detection and response to emerging infectious diseases (EIDs).

The integration of advanced technologies, such as environmental DNA (eDNA) analysis, artificial intelligence-driven diagnostics, and remote sensing, has revolutionized wildlife disease monitoring, enabling more precise and rapid pathogen detection. By leveraging these innovations alongside traditional field surveillance methods, *Conservation Medicine* can adopt a more proactive stance in addressing the root causes of disease emergence. This approach not only reduces the risk of future pandemics but also safeguards biodiversity conservation by preserving ecosystem integrity and promoting sustainable coexistence between humans and wildlife.

As global existential health threats continue to evolve, investment in interdisciplinary research, policy development, and community-based disease surveillance, monitoring, and rapid response programs will be crucial in building resilient disease monitoring systems. By prioritizing wildlife disease surveillance as a critical component of *Conservation Medicine*, stakeholders can strengthen our collective capacity to anticipate, prevent, and respond to emerging health challenges while fostering a healthier planet for all species.

5.3 Disease Spillover Prevention in Conservation Medicine.

5.3.1 Introduction.

Disease spillover, which refers to the transmission of pathogens from wildlife to humans or domestic animals, presents a profound challenge with far-reaching consequences. Such events threaten the public's health by facilitating the emergence of zoonotic diseases, destabilize biodiversity conservation by disrupting ecosystem services, and impose substantial economic burdens on global economies through healthcare costs, trade restrictions, and productivity losses. High-profile examples, including the spread of SARS-CoV-2, underscore the urgency of addressing this issue at its root.

Conservation Medicine, previously defined, is an interdisciplinary field that integrates human, animal, and ecosystem health, plays a pivotal role in mitigating spillover risks. By examining the intricate interconnections among species and their shared ecosystems, *Conservation Medicine* provides a holistic framework for identifying and addressing the drivers of disease emergence.

Preventing spillover requires a comprehensive, complex approach that prioritizes several key strategies:

1. *Habitat Conservation*:
   Protecting and restoring natural habitats plays a vital role in minimizing human-wildlife interactions, a key factor in preventing disease spillover. Maintaining the integrity of ecosystems helps to stabilize the services they provide, reducing the chances of wildlife being forced into closer contact with human communities or domestic animals.

2. *Enhanced Disease Surveillance, Monitoring, and Rapid Response Systems*: Continuous monitoring of wildlife health is essential for the early detection of zoonotic pathogens. Advanced tools, such as genomic sequencing and real-time data analytics, can improve our ability to track and predict potential spillover events, enabling timely interventions.

3. Biosecurity Measures:
Strengthening biosecurity practices in agriculture, wildlife trade, and tourism can curb the transmission of pathogens at the human-animal interface. This includes implementing rigorous health protocols, regulating wildlife markets, and monitoring livestock health to prevent cross-species transmission.

4. *Community Engagement*:
Involving local communities and indigenous populations in spillover prevention initiatives fosters sustainable solutions. Their participation enhances awareness, supports conservation efforts, and integrates traditional knowledge with modern scientific approaches to manage disease risks effectively.

By adopting these strategies within the *Conservation Medicine* framework, the frequency and impact of spillover events may be significantly reduced, safeguarding not only the public's health but also the delicate balance of our planet's ecosystems. This holistic and integrated approach emphasizes that the health of humans, animals, and the environment are inseparable, reinforcing the critical importance of collaboration across disciplines and sectors.

5.3.2 Habitat Conservation and Biodiversity Protection.

Preserving natural ecosystems is a cornerstone strategy for preventing disease spillover events. The degradation of ecosystem's services through habitat destruction, deforestation, and land-use changes drives increased interactions between humans and wildlife, creating pathways for pathogen transmission (Keesing et al., 2010). These disruptions often force wildlife into closer proximity with human populations and domestic animals, elevating the risk of zoonotic spillover.

Biodiversity conservation plays a crucial role in mitigating these risks, particularly through the *dilution effect*, which suggests that ecosystems with high species diversity can limit pathogen amplification in specific reservoir hosts. A reduction in biodiversity conservation diminishes this protective effect, thereby increasing the likelihood of pathogens spreading unchecked among reservoir species and spilling over to humans or livestock (Ostfeld & Keesing, 2012).

Conservation efforts aimed at safeguarding and restoring ecosystem balance are vital for reducing spillover risks. By maintaining intact habitats and minimizing human encroachment into wildlife territories, these initiatives help stabilize ecosystem's services, reduce stress on wildlife populations, and mitigate the conditions that favor pathogen emergence. Additionally, such efforts support other critical benefits, including climate regulation, water purification, and the overall health of ecosystems, creating a more resilient environment for all species.

Integrating habitat conservation with broader public health and biodiversity conservation policies—within frameworks such as *One Health*—offers a proactive approach to addressing the interconnected challenges of disease spillover, biodiversity loss, and human well-being. This strategy underscores the importance of preserving our natural world as a safeguard for global health and ecosystem integrity.

5.3.3 Disease Surveillance and Early Detection.

Effective disease surveillance and early detection systems are foundational to identifying emerging pathogens and preventing their spread across human populations, wildlife, and domestic animals. Early detection enables timely interventions, reducing the risk of outbreaks escalating into global health crises. The implementation of the *One Health* approach, which emphasizes the inextricable interconnectedness of human, animal, and environmental health, further strengthens surveillance efforts. By facilitating collaboration across these sectors, *One Health* improves the accuracy of outbreak prediction and enhances the development of coordinated responses (Murray et al., 2022).

Sentinel species play a pivotal role in zoonotic disease surveillance. These organisms, which are highly susceptible to specific infections, act as early warning indicators for potential threats. Monitoring the health of sentinel species allows scientists to detect pathogens before they reach humans or domestic animals, providing crucial lead time for intervention (Plowright et al., 2017).

Technological advancements have revolutionized the capabilities of disease surveillance systems, enabling more sophisticated methods for pathogen detection. Metagenomics, a powerful tool that analyzes genetic material from environmental samples, allows researchers to identify a wide array of pathogens, including those previously unknown. Similarly, environmental DNA (eDNA) analysis has emerged as a transformative scientific approach, capable of detecting trace amounts of genetic material shed by organisms into their surroundings. These technologies enhance our ability to identify pathogens before they spill over to new hosts, allowing for proactive measures to mitigate risks (Carroll et al., 2018).

By integrating robust disease surveillance and early detection systems, sentinel species monitoring, and cutting-edge technologies within the *One Health* framework, there is enhanced anticipation, response, and ultimately prevention of zoonotic disease outbreaks. These efforts not only protect the public's health but also contribute to the preservation of biodiversity conservation and ecosystem resilience.

5.3.4 Biosecurity and Risk Mitigation Measures in Disease Spillover Prevention.

Implementing strict biosecurity measures is essential for limiting the transmission of pathogens across wildlife, livestock, and human populations. These measures play a pivotal role in reducing the risk of zoonotic disease emergence by addressing key interfaces where cross-species transmission is most likely to occur.

One significant strategy involves regulating the wildlife trade, which is often associated with the introduction and spread of zoonotic pathogens. Enforcing stricter controls, such as banning the sale of live animals in wet markets, can reduce opportunities for pathogen spillover from wildlife to humans or domestic animals (Karesh et al., 2012). Additionally, enhancing veterinary controls—through rigorous health inspections, quarantine protocols, and the monitoring of wildlife health—provides an added layer of protection against the spread of infectious diseases.

Improving livestock management practices is another critical component of effective biosecurity. Controlled grazing systems, for example, help reduce contact between domestic animals and potentially infected wildlife, thereby minimizing opportunities for cross-species transmission. Vaccination programs for livestock serve as a proactive measure to prevent these animals from becoming intermediate hosts for zoonotic diseases, further breaking the chain of transmission (Jones et al., 2013).

Moreover, technological advancements such as real-time tracking of animal movements, automated health monitoring systems, and predictive modeling tools enhance biosecurity practices by identifying risks and vulnerabilities with greater precision. Integrating these innovations within farming and wildlife management systems can lead to more robust disease prevention strategies.

Lastly, community engagement and education are crucial to ensuring compliance with biosecurity protocols. By raising awareness about the importance of these measures and their role in safeguarding the public's health, local populations, farmers, and policymakers can work together to build a more resilient system of disease prevention.

Incorporating these multi-faceted biosecurity measures within broader *One Health* frameworks creates a comprehensive and sustainable approach to mitigating zoonotic disease risks, benefiting human, animal, and environmental health alike.

5.3.5 Community Engagement and Education in Disease Spillover Prevention.

Involving local communities in conservation and public health initiatives is a crucial step in developing effective, sustainable strategies to prevent zoonotic spillover events. Community engagement fosters a sense of empowered ownership and collective responsibility for health and environmental stewardship, bridging gaps between scientific research, policymaking, and on-the-ground practices.

Traditional Ecological Knowledge (TEK) represents a vital asset in this effort. The deep understanding that indigenous and local populations have of their surrounding ecosystems and the services they provide are invaluable insights into wildlife behavior, habitat changes, and the patterns of disease emergence. When combined with participatory epidemiology, which actively involves communities in disease monitoring and reporting, these approaches empower local populations to identify and respond to health threats effectively (Bonwitt et al., 2018). Such integration of local knowledge with scientific methodologies strengthens both early warning systems and response strategies.

Outreach programs tailored to the unique needs and contexts of communities are another essential component of spillover prevention. These programs promote practices that minimize direct human exposure to high-risk species, such as safe hunting techniques and proper wildlife handling protocols. For example, educating hunters on the risks of zoonotic pathogens and the importance of avoiding contact with potentially infected animals can reduce spillover risks. Programs also emphasize alternative livelihood opportunities, such as ecotourism, sustainable agriculture, or aquaculture, which not only enhance economic stability but also reduce dependence on activities like wildlife trade or bushmeat consumption, known to be high-risk behaviors (Daszak et al., 2020).

Moreover, community-based initiatives are most effective when coupled with broader policy frameworks, such as *One Health* or *EcoHealth*, which advocate for interdisciplinary collaboration among conservationists, public health professionals, veterinarians, physicians, and local leaders and stakeholders. When communities are fully engaged and supported, they become essential partners in safeguarding both biodiversity conservation and human health.

By embedding community knowledge, practices, and aspirations into spillover prevention strategies, we create solutions that are not only scientifically sound but also socially equitable and culturally resonant. These approaches highlight the profound interconnectedness of human well-being, wildlife health, and ecosystem integrity.

5.3.6 Conclusion.

Preventing disease spillover demands a comprehensive and interdisciplinary approach that integrates conservation strategies, technological innovations, and public health interventions. By addressing the complex interactions between wildlife, humans, and domestic animals, *Conservation Medicine* provides a framework for mitigating zoonotic threats at their source. This requires proactive efforts to preserve natural ecosystems, reduce human-wildlife conflict, and implement sustainable land-use practices that minimize pathogen transmission risks.

Advanced technological tools, such as genomic sequencing, remote sensing, and artificial intelligence-driven disease modeling, play a pivotal role in enhancing disease surveillance and early detection of emerging infectious diseases (EIDs). These innovations facilitate real-time monitoring of wildlife health, enabling rapid response measures to prevent spillover events before they escalate into pandemics. Additionally, public health initiatives that incorporate *One Health* principles—recognizing the inextricable interconnectedness of human, animal, and environmental health—ensure that disease prevention strategies are both effective and sustainable.

Global collaboration and interdisciplinary and multisectoral partnerships are crucial in addressing the root causes of zoonotic spillover. Scientists, veterinarians, physicians, policymakers, and other local stakeholders must work together to implement empirically-driven evidence-based interventions, strengthen biosecurity measures, and regulate high-risk activities such as wildlife trade and deforestation. Investing in capacity-building efforts, such as training and developing frontline healthcare workers and empowering indigenous and rural populations, further enhances resilience against emerging infectious diseases.

Ultimately, preventing future pandemics and fostering a sustainable coexistence with wildlife requires a shift toward a more integrated and preventive approach. By prioritizing ecosystem health, leveraging technological advancements, and fostering strong partnerships across disciplines, humanity can reduce the likelihood of disease spillover and create a healthier, more balanced relationship with the natural world.

5.4 Role of Protected Areas and Wildlife Corridors in Biodiversity Conservation.

5.4.1 Introduction.

Biodiversity conservation is vital for preserving the intricate balance of ecosystems, supporting their stability, adaptability, and resilience in the face of environmental changes. A diverse range of species contributes to critical ecosystem services that are indispensable to human survival, including the purification of air and water, the maintenance of soil health, and the production of food. Furthermore, biodiversity conservation enhances the capacity of ecosystems to recover from disturbances and adapt to climate variability, thereby ensuring their long-term functionality.

Protected areas (PAs) and wildlife corridors are integral components of effective biodiversity conservation strategies. PAs serve as sanctuaries where ecosystems can thrive without excessive human interference, fostering the preservation of habitats and protecting species from extinction. Wildlife corridors, on the other hand, promote ecosystem connectivity by linking fragmented habitats, allowing the natural movement of species and genetic exchange. Together, these strategies not only counteract the adverse effects of habitat fragmentation but also help mitigate the impacts of human-induced climate change and encroachment.

In addition to safeguarding the survival of individual species, these conservation efforts play a pivotal role in sustaining broader ecosystem's processes, such as pollination, nutrient cycling, and climate regulation. By prioritizing the protection and restoration of biodiversity conservation through initiatives like PAs and wildlife corridors, humanity invests in the health and resilience of the planet, ultimately securing a sustainable future for all life forms.

5.4.2 Protected Areas as Biodiversity Conservation Refuges.

Protected areas (PAs) are indispensable tools in the global effort to protect biodiversity conservation, providing secure habitats for threatened and endangered species while shielding them from deforestation, poaching, and habitat destruction. These sanctuaries offer both legal and physical protections that help maintain the ecosystem balance required for species survival and ecosystem stability. Studies have demonstrated that well-managed PAs significantly mitigate biodiversity loss by preserving habitat integrity, supporting stable species populations, and promoting ecosystem services essential to life on Earth (Watson et al., 2014). Globally, the creation of national parks, wildlife reserves, and marine protected areas has proven effective in slowing biodiversity conservation declines by curbing harmful human activities, fostering sustainable land use practices, and enhancing critical ecosystem services such as pollination, water purification, and climate regulation (Dudley et al., 2010).

Despite these benefits, the success of PAs depends heavily on robust funding, sound governance, and effective enforcement mechanisms. Challenges such as illegal logging, agricultural expansion, encroachment, and insufficient law enforcement jeopardize the ability of PAs to fulfill their conservation objectives. Without proactive measures to address these issues, protected areas risk falling short of their potential to halt biodiversity loss and ensure the survival of countless species.

To maximize the impact of protected areas in biodiversity conservation, it is essential to adopt multidimensional strategies. Strengthening management frameworks by implementing advanced monitoring and rapid response systems and employing skilled personnel can enhance the effectiveness of PAs. Engaging and empowering local communities in biodiversity conservation efforts not only fosters environmental stewardship but also ensures that sustainable practices align with their socioeconomic needs. Furthermore, securing long-term financial sustainability—through innovative funding mechanisms such as eco-tourism, public-private partnerships, and international conservation grants—can provide the resources needed for the continual upkeep and improvement of protected areas.

By addressing these challenges and fortifying biodiversity conservation strategies, protected areas can fulfill their role as critical bastions of biodiversity, ensuring the preservation of our planet's invaluable ecological heritage for future generations.

5.4.3 Wildlife Corridors and Ecosystem Connectivity.

Wildlife corridors are strategically designated land areas that play a critical role in connecting fragmented habitats, enabling species to move freely across landscapes for migration, dispersal, and reproduction. By facilitating these natural movements, corridors help maintain genetic diversity within populations, thereby reducing the risks associated with inbreeding depression and enhancing the overall resilience of species to environmental challenges (Hilty et al., 2020). These natural passageways allow ecosystems to function more cohesively, mitigating the negative consequences of habitat fragmentation caused by urban development, deforestation, and other human activities.

One exemplary success story is the Yellowstone-to-Yukon (Y2Y) corridor in North America, which spans an expansive region to ensure connectivity for wildlife species such as wolves, bears, and elk. This corridor supports not only the survival of individual populations but also the broader ecosystem services that sustain them, thereby enhancing resilience across vast landscapes (Noss et al., 2012). Similarly, the Central Amazon Corridor in Brazil plays a crucial role in linking rainforest ecosystems. By maintaining connectivity, it provides safe passage for iconic species like jaguars and primates while simultaneously supporting the region's unparalleled biodiversity (Peres et al., 2018).

Beyond facilitating day-to-day ecosystem's functions, wildlife corridors are increasingly recognized as vital climate adaptation strategies. As global temperatures rise and habitat conditions shift, these corridors allow species to adjust their ranges in search of suitable environments. By maintaining pathways for movement, they enable ecosystems to adapt more effectively to the dynamic and unpredictable patterns brought about by climate change (Keeley et al., 2018).

The significance of wildlife corridors extends far beyond individual species, as they safeguard broader ecosystem integrity and contribute to the provision of essential ecosystem services. By mitigating habitat fragmentation and fostering connectivity, they ensure the long-term survival of species and ecosystems alike, representing a cornerstone of modern conservation efforts. Prioritizing their creation, expansion, and management is imperative for building a resilient and biodiverse future.

5.4.4 Challenges and Future Directions for Protected Areas and Wildlife Corridors.

Despite their undeniable importance, both protected areas (PAs) and wildlife corridors remain vulnerable to a range of persistent threats, including land-use changes, infrastructure development, and insufficient policy enforcement. Urbanization, agricultural expansion, and resource extraction often encroach upon these conservation zones, undermining their ability to safeguard biodiversity. Moreover, weak governance and inadequate regulatory frameworks exacerbate challenges, leaving many PAs and corridors unable to realize their full potential as conservation tools.

To ensure the long-term success of these initiatives, conservation efforts must adopt a holistic, interdisciplinary and multisectoral approach. Integrating scientific research is vital for generating empirically-derived evidence-based strategies to address biodiversity loss and monitor the effectiveness of interventions. Stakeholder engagement—spanning governments, non-governmental organizations, private sectors, and local communities—creates collaborative frameworks that align conservation goals with socioeconomic priorities. Developing and implementing robust policy frameworks can further provide the legal and institutional backbone necessary for protecting and maintaining these critical areas.

Strengthening transboundary conservation initiatives is particularly important in regions where ecosystems and wildlife populations transcend national borders. Collaborative efforts between neighboring countries can enhance ecological connectivity and ensure the continuity of vital habitats. Enhancing ecosystem restoration projects also plays a key role in rejuvenating degraded environments, rebuilding habitats, and supporting species populations. Furthermore, incorporating Indigenous Knowledge (IK) offers invaluable insights into sustainable land stewardship, resource management, and conservation practices honed over centuries (Garnett et al., 2018).

Technological advancements offer powerful tools for overcoming biodiversity conservation challenges and improving planning processes. Satellite monitoring and Geographic Information Systems (GIS) mapping enable accurate tracking of habitat changes, while artificial intelligence and computational science supports the analysis of complex data to identify priority areas for protection and optimize corridor design (Díaz et al., 2019). These technologies can stream-line biodiversity conservation efforts, ensuring resources are strategi-cally allocated for maximum impact.

Sustaining and expanding biodiversity conservation efforts requires significant investments in funding, underscoring the impor-tance of innovative financing mechanisms such as public-private partnerships and international conservation grants. Global coopera-tion is equally critical, fostering unity and collective action to tackle biodiversity loss on a planetary scale. By leveraging scientific insights, technological tools, and community-driven approaches, conserva-tion initiatives can thrive, protecting the natural world and securing a sustainable future for all species.

5.4.5 Conclusion.

Protected areas and wildlife corridors are essential pillars of biodiver-sity conservation, serving as sanctuaries for species, preserving ecosys-tems, and maintaining ecological connectivity. By safeguarding crit-ical habitats, these conservation tools help mitigate biodiversity loss, support species adaptation to environmental changes, and sustain ecosystem services that benefit both wildlife and human populations. However, their effectiveness depends on integrated conservation plan-ning, strong policy frameworks, and active community participation.

As biodiversity conservation faces unprecedented threats from habitat destruction, climate change, and human encroachment, the role of protected areas and corridors has become increasingly vital. Expanding these conservation efforts requires strategic land-use planning, ensuring that protected areas are not only established but also effectively managed and enforced. This involves securing legal protections, preventing illegal activities such as poaching and defor-estation, and enhancing habitat restoration initiatives. Additionally, incorporating science-based approaches, such as ecological modeling and landscape connectivity assessments, can optimize the design and placement of new protected areas and corridors to maximize their conservation impact.

Policy support is equally crucial in strengthening the resilience of conservation areas. Governments, conservation organizations, and international bodies must work collaboratively to implement and uphold policies that prioritize habitat protection and prevent environmentally harmful development. Incentivizing sustainable land-use practices, integrating conservation goals into national planning, and ensuring adequate funding for conservation programs are necessary steps toward long-term biodiversity preservation. Moreover, transboundary conservation initiatives can enhance ecological connectivity across political borders, fostering cooperation among nations to protect shared ecosystems and migratory species.

Community involvement plays a critical role in the success of conservation efforts. Engaging local and Indigenous communities in conservation planning and decision-making can enhance the effectiveness and sustainability of protected areas. *Traditional Ecological Knowledge* (TEK), when combined with scientific research, offers valuable insights into habitat management and species conservation. Programs that provide economic incentives, such as ecotourism and sustainable resource use, can also strengthen local support for conservation by demonstrating the tangible benefits of biodiversity preservation.

Given the accelerating pace of biodiversity loss, expanding and reinforcing protected areas and wildlife corridors is no longer an option but an urgent necessity. Investments in conservation research, technological innovations such as remote sensing and artificial intelligence for habitat monitoring, and multi-sectoral collaboration will be essential in ensuring the long-term success of these strategies. By prioritizing habitat conservation and fostering a global commitment to biodiversity conservation, we can work toward a future where ecosystems remain intact, species thrive, and humanity coexists harmoniously with nature.

5.5 Conservation Medicine Interventions: Vaccination Programs, Habitat Restoration, and Captive Breeding.

5.5.1 Introduction.

Conservation Medicine interventions are integral to both preserving biodiversity and addressing the numerous threats that jeopardize the survival of endangered species. By intertwining principles of ecosystem and environmental health, veterinary science, and environmental stewardship and sustainability, these interventions provide innovative solutions to the complex challenges facing wildlife and their habitats.

One of the most impactful strategies in *Conservation Medicine* is the implementation of *vaccination programs.* These initiatives aim to control and prevent disease outbreaks in wild animal populations, which can often have devastating ripple effects on ecosystems. For example, targeted vaccination efforts have been successful in curbing diseases like canine distemper and rabies, both of which threaten not only wildlife but also human and domestic animal health in shared ecosystems.

Another cornerstone of *Conservation Medicine* is *habitat restoration.* This involves rehabilitating degraded ecosystems to reestablish the intricate web of life that supports biodiversity conservation. Restoration efforts may include reforestation, wetland recovery, or invasive species removal, all of which create environments where native flora and fauna can both survive and thrive. Such efforts not only benefit individual species but also contribute to ecosystem resilience, ensuring the stability of critical environmental functions like carbon sequestration and water purification.

Equally significant are *captive breeding programs*, which focus on the recovery and long-term survival of critically endangered species. By carefully managing breeding in controlled environments, these programs help bolster population numbers and preserve genetic diversity. Captive-bred individuals are often reintroduced into the wild as part of broader recovery plans, enhancing the ecosystem balance and restoring species that may otherwise have been lost forever.

These interventions do not operate in isolation; they are typically integrated into comprehensive conservation strategies that address the complex nature of biodiversity loss. Beyond stabilizing ecosystems and preventing extinctions, such approaches underscore the inextricable interconnectedness of human, animal, and environmental health—a perspective central to the principles of *Conservation Medicine*. By addressing the root causes of ecosystem degradation and implementing evidence-based solutions, *Conservation Medicine* plays a vital role in shaping a sustainable future for both wildlife and humanity.

5.5.2 Vaccination Programs in Wildlife Conservation.

Vaccination programs are an essential disease management tool in *Conservation Medicine*, particularly for species threatened by infectious diseases. Wildlife populations are vulnerable to both naturally occurring diseases and those transmitted by domestic animals. Immunization efforts have helped prevent devastating population declines in several species.

One of the most successful examples of vaccination in *Conservation Medicine* is the oral rabies vaccination of wild carnivores. Programs targeting Ethiopian wolves *(Canis simensis)* and black-footed ferrets *(Mustela nigripes)* have significantly reduced disease transmission and improved population stability (Knobel et al., 2008; Randall et al., 2020). Similarly, vaccination efforts against canine distemper virus (CDV) have protected endangered species such as the Amur tiger *(Panthera tigris altaica)* and African wild dogs *(Lycaon pictus)* from catastrophic outbreaks (Gilbert et al., 2020).

Although vaccination programs have demonstrated significant success in safeguarding wildlife health, several challenges persist, highlighting the complexities of implementing such interventions on a large scale. Among these challenges are the logistical difficulties involved in delivering vaccines to wild animal populations. Unlike domesticated animals, wildlife often inhabit remote, vast, and densely vegetated areas, making it labor-intensive and costly to locate and immunize them effectively. Additionally, the process of capturing or handling animals for vaccination can introduce stress and potential harm to the individuals and disrupt their natural behaviors.

There are also concerns about the potential ecosystem and environmental consequences of introducing vaccines into wild populations. For instance, vaccine administration might inadvertently alter predator-prey dynamics or interfere with natural selection processes, leading to unforeseen ecosystem imbalances. Similarly, improper vaccine deployment or dosage might carry risks of contamination or adverse side effects, raising ethical and environmental considerations.

Despite these obstacles, promising advancements in vaccination technology are paving the way for more feasible and effective solutions. For example, the development of oral bait vaccines has revolutionized wildlife immunization strategies by enabling the distribution of vaccines in a manner that minimizes human-wildlife interaction (Maki et al., 2017). These vaccines can be strategically placed in habitats where they are likely to be consumed by target species, providing a practical and scalable solution to the challenges of direct administration.

Furthermore, breakthroughs in genetically engineered immunization methods are enhancing the precision and efficacy of wildlife vaccination efforts. These innovations include the use of recombinant vaccines, which can be tailored to target specific pathogens while reducing the risks of adverse effects (Sarker et al., 2019). Such methods hold immense potential for addressing complex wildlife diseases, such as those caused by emerging zoonotic pathogens (e.g., SARS-CoV-2, Ebola, etc.).

Overall, while challenges remain inherent to wildlife vaccination programs, continuous technological and methodological advancements are steadily overcoming these hurdles. By leveraging tools like oral bait vaccines and cutting-edge genetic engineering, conservationists are making strides toward protecting vulnerable wildlife populations and preserving ecosystem health in a sustainable and effective manner.

5.5.3 Habitat Restoration as a Conservation Medicine Strategy.

Habitat loss is one of the primary drivers of biodiversity conservation decline, and habitat restoration aims to reverse this trend by rehabilitating degraded ecosystems. Restoring ecosystems not only benefits wildlife populations but also enhances ecosystem services such as carbon sequestration, water filtration, and climate resilience.

Successful habitat restoration projects include reforestation efforts in the Atlantic Forest of Brazil, which have helped recover critical habitat for endangered species like the golden lion tamarin *(Leontopithecus rosalia)* (Banks-Leite et al., 2014). Similarly, wetland restoration in North America has improved breeding success for migratory bird populations and provided flood mitigation benefits (Zedler & Kercher, 2005).

One of the most ambitious habitat restoration projects is the *Great Green Wall Initiative* in Africa, which aims to combat desertification by restoring 100 million hectares of degraded land across the Sahel region. This initiative not only helps biodiversity conservation but also strengthens food security and climate adaptation for local communities (Berrahmouni et al., 2016).

Despite its benefits, habitat restoration faces challenges, including the high cost of reforestation projects, the slow recovery rate of some ecosystems, and the need for long-term maintenance. Integrating community-led biodiversity conservation and sustainable land-use practices is key to ensuring the lasting success of restoration efforts.

5.5.4 Captive Breeding and Species Recovery Programs.

Captive breeding programs are used to prevent the extinction of critically endangered species by breeding individuals in controlled environments before reintroducing them into the wild. These programs have played a vital role in saving species from extinction, often serving as a last resort for populations with dangerously low numbers.

One of the most well-known successes in captive breeding is the recovery of the California condor *(Gymnogyps californianus)*, which was reduced to only 27 individuals in the 1980s. Through intensive captive breeding and reintroduction efforts, the population has grown to over 500 birds, with more than half now living in the wild (Ralls & Ballou, 2020).

Similarly, the Arabian oryx *(Oryx leucoryx)*, once extinct in the wild, was successfully reintroduced through captive breeding efforts led by conservation organizations in the Middle East. The species now thrives in protected areas, demonstrating the potential of breeding programs to restore once-lost species to their native habitats (Stanley Price, 2018).

However, captive breeding programs must be carefully managed to avoid genetic bottlenecks, behavioral changes that hinder reintroduction, and over-reliance on human intervention. Advances in genetic diversity management and the use of conservation genomics have improved the effectiveness of captive breeding efforts, ensuring that reintroduced populations remain viable in the long term (Frankham et al., 2017).

5.5.5 Conclusion.

Conservation interventions such as vaccination programs, habitat restoration, and captive breeding are vital components of global efforts to protect biodiversity conservation and prevent species extinction. These approaches help mitigate the devastating impacts of habitat loss, man-made climate change, and emerging infectious diseases (EIDs), which are among the leading threats to wildlife populations. By employing a combination of disease control, ecosystem rehabilitation, and species recovery programs, *Conservation Medicine* interventions can improve the resilience of ecosystems and ensure the long-term survival of vulnerable species.

Each of these conservation strategies comes with its own unique challenges. Vaccination programs, while effective in controlling disease outbreaks among endangered species, require logistical coordination, ongoing research to adapt to evolving pathogens, and ethical considerations regarding human intervention in wildlife populations. Habitat restoration efforts, though crucial for reversing environmental degradation, can be resource-intensive and require long-term monitoring to ensure that restored ecosystems support biodiversity conservation. Captive breeding, often a last resort for critically endangered species, must carefully manage genetic diversity and prevent the loss of natural behaviors that are essential for species reintroduction into the wild.

Despite these challenges, advancements in technology, strong policy support, and community involvement are continuously improving the effectiveness of conservation interventions. Emerging tools such as genomic sequencing enhance captive breeding programs by preventing genetic bottlenecks, while remote sensing and artificial intelligence facilitate large-scale habitat restoration and biodiversity monitoring. Additionally, policy frameworks such as the *Convention on Biological Diversity (CBD)* and the *Endangered Species Act (ESA)* provide legal protections and funding mechanisms that support conservation efforts worldwide.

Community engagement and empowerment remains a critical factor in the success of biodiversity conservation initiatives. Indigenous knowledge and local environmental stewardship play a key role in sustainable habitat restoration, while community-led wildlife monitoring programs enhance the reach and efficiency of conservation efforts. Public awareness campaigns and ecotourism initiatives also contribute to long-term conservation by creating economic incentives that encourage habitat protection and wildlife preservation.

As global biodiversity conservation faces increasing threats, the need for integrated, empirically-driven evidence-based *Conservation Medicine* strategies and interventions has never been more urgent. Expanding and refining vaccination programs, habitat restoration

195

projects, and captive breeding efforts will be essential to safeguarding ecosystems and protecting the world's most vulnerable species. Looking ahead, the integration of environmental and ecosystem science, cutting-edge technological advancements, and community-led initiatives in biodiversity conservation will be vital in securing the sustainability and impact of these efforts for generations to come. By fostering collaboration across disciplines and empowering local communities, this approach can address complex ecological challenges while maintaining its prominence as a global priority.

5.6 Community Engagement and Sustainable Development Initiatives in Conservation Medicine.

5.6.1 Introduction.

Conservation Medicine, as previously discussed, is an interdisciplinary field that recognizes the inextricable interconnected health of humans, animals, and ecosystems. Effective conservation efforts require the integration of community engagement and sustainable development initiatives to ensure long-term success. Local communities play a crucial role in preserving biodiversity conservation, mitigating zoonotic disease risks, and maintaining ecosystem services. By fostering sustainable livelihoods, promoting traditional environmental health and ecosystems' knowledge, and involving communities in decision-making processes, *Conservation Medicine* can achieve both environmental and public health goals.

5.6.2 The Role of Community Engagement in Conservation Medicine.

Community engagement plays a pivotal role in addressing biodiversity conservation challenges, as local populations frequently act as the primary custodians of their surrounding environment. By involving communities in conservation initiatives, not only are local stakeholders empowered to take ownership of these efforts, but they also gain a heightened awareness of ecological issues and adopt more sustainable resource management practices (Berkes, 2018). This participatory approach ensures that conservation strategies resonate with the needs, preferences, and cultural priorities of the communities involved, thereby increasing the likelihood of long-term success and resilience.

One prominent example of this approach is the *Community-Based Natural Resource Management (CBNRM)* model, which has been employed across diverse regions, particularly in Africa and Asia. This model emphasizes local leadership and governance, enabling communities to take an active role in managing natural resources and ecosystem services. In Namibia, for instance, communal conservancies have successfully empowered local populations to oversee wildlife resources within their territories. These conservancies have not only facilitated significant biodiversity recovery, such as increasing populations of elephants and other iconic species, but they have also improved livelihoods by generating income through eco-tourism, sustainable hunting quotas, and other wildlife-related enterprises (Naidoo et al., 2016).

Beyond natural resource and ecosystem services management, community participation is also proving invaluable in advancing the goals of *Conservation Medicine*, particularly in the realms of disease surveillance and prevention. The adoption of participatory epidemiology exemplifies this synergy. In regions prone to zoonotic spillover—where diseases jump from animals to humans—local communities are trained to observe and report disease symptoms in wildlife and livestock. This early detection system enables swift interventions, such as quarantining infected areas, thereby reducing the risks of disease transmission and potential pandemics. For example, community-based training initiatives have been instrumental in monitoring diseases like Ebola and anthrax in vulnerable regions, strengthening both public health systems and wildlife conservation outcomes (Bonwitt et al., 2018).

Ultimately, fostering strong community involvement in bio-diversity conservation efforts enhances not only ecosystem service's outcomes but also social and economic benefits, creating a virtu-ous cycle of sustainability. By integrating local knowledge, cultural values, and scientific expertise, community-driven approaches can bridge the gap between biodiversity conservation and human devel-opment, ensuring a harmonious balance for future generations.

5.6.3 Sustainable Development Initiatives in Conservation Medicine.

Sustainable development initiatives help ensure that conservation efforts are not at odds with the economic needs of local communi-ties. *Conservation Medicine* recognizes that environmental protection must be integrated with poverty reduction, food security, and health equity (McDermott et al., 2019).

1. *Sustainable Livelihoods and Ecotourism.*
   Sustainable livelihood programs provide alternative income sources that reduce dependence on environmentally destruc-tive activities such as poaching and deforestation. Ecotourism, when managed responsibly, generates revenue for biodiversity conservation projects while offering employment opportunities to local communities. For example, in Rwanda, gorilla ecotour-ism has contributed to both wildlife conservation and economic development, with a portion of tourism revenue reinvested into local healthcare and education (Spenceley et al., 2017).

2. *Sustainable Agriculture and Agroforestry.*
   Agricultural expansion is a leading driver of habitat loss and bio-diversity decline. *Conservation Medicine* promotes sustainable agricultural practices, such as agroforestry, to enhance food pro-duction while preserving biodiversity. In the Amazon, agrofor-estry initiatives that integrate native tree species with crops have

reduced deforestation rates and improved soil health (Schroth & McNeely, 2011). These practices support both environmental and human health by maintaining ecosystem services such as water filtration and pollination.

3. *Wildlife-Friendly Disease Prevention Strategies.*
Community-led health initiatives that focus on reducing human-wildlife conflict and improving biosecurity help prevent zoonotic spillover events. For instance, vaccination programs for livestock in East Africa have been used to protect both domestic animals and local wildlife populations from infectious diseases such as rabies and rinderpest (Groot et al., 2020). These efforts align with the *One Health* approach, which integrates human, animal, and environmental health to address complex conservation challenges.

5.6.4 Challenges and Future Directions in Conservation Medicine.

While community engagement and sustainable development initiatives present promising pathways for addressing biodiversity conservation and environmental and ecosystem challenges, they are not without obstacles. Significant hurdles include funding limitations, which constrain the implementation and scalability of community-driven programs; policy barriers, such as conflicting regulations or insufficient legal frameworks, that hinder effective action; and socio-political conflicts, including disputes over land use, resource allocation, and governance structures, which can undermine collaborative efforts (Pretty et al., 2009).

Ensuring the long-term success of these initiatives requires a robust commitment to continuous investment in key areas. Building the capacity of local communities is essential, equipping them with the knowledge, skills, and resources needed to engage meaningfully in conservation efforts. Additionally, fostering knowledge exchange between stakeholders—such as researchers, practitioners, and community members—creates opportunities for sharing innovative ideas and best practices. Finally, equitable decision-making processes must be prioritized, ensuring that all voices, particularly those of marginalized groups, are heard and valued in conservation planning and implementation.

To overcome challenges and maximize impact, future conservation efforts should focus on the following strategic priorities:

1. *Strengthening local leadership and governance.*
    Empowering communities to take an active role in conservation decision-making is crucial for fostering ownership and accountability. This involves providing technical support, enhancing leadership skills, and developing transparent governance mechanisms that allow communities to manage natural resources effectively.

2. *Enhancing interdisciplinary collaborations.*
    Addressing the complex and interconnected nature of biodiversity conservation requires collaboration across disciplines. Conservationists, public health experts, policymakers, economists, and other stakeholders must work together to integrate environmental, social, and economic perspectives into holistic and sustainable solutions.

3. *Expanding financial incentives.*
    Initiatives such as Payment for Ecosystem Services (PES) can play a transformational role in ensuring sustainable community participation. PES programs compensate local populations for maintaining ecosystem services, such as forest conservation, water quality protection, and carbon sequestration. By linking conservation outcomes to tangible economic benefits, these incentives encourage long-term commitment and create win-win scenarios for both biodiversity and human well-being.

Ultimately, by addressing existing barriers and prioritizing these areas, biodiversity conservation efforts can foster resilient, inclusive, and sustainable systems that effectively balance ecosystem preservation with community development. This integrated approach will be critical in building a future where humans and nature can thrive together.

5.6.5 Conclusion.

Integrating community engagement and sustainable development into *Conservation Medicine* strengthens both ecosystem resilience and human wellness and well-being, ensuring that biodiversity conservation efforts are not only effective but also socially just and economically viable. Conservation strategies that incorporate local knowledge, aligned economic incentives, and participatory leadership and governance foster long-term commitment from communities, increasing the likelihood of successful biodiversity preservation, disease prevention, health protection, and health promotion.

By aligning biodiversity conservation goals with local economic and health needs, initiatives such as sustainable agriculture, eco-tourism, and community-based wildlife monitoring create mutually beneficial outcomes for both biodiversity conservation and people. When communities perceive direct benefits—such as improved food security, higher-quality healthcare services, and resilient livelihoods—they become active participants in biodiversity conservation efforts rather than passive stakeholders. For instance, payment for ecosystem services (PES) programs provide financial incentives to landowners and farmers for preserving forests and watersheds, linking economic stability with environmental stewardship. Similarly, sustainable livestock vaccination programs not only protect endangered wildlife from zoonotic diseases but also safeguard community livelihoods by reducing livestock losses.

Moving forward, strengthening partnerships among local communities, conservation scientists, public health experts, and policymakers will be essential to ensuring that biodiversity conservation efforts remain both scientifically thorough and socially equitable. Collaborative decision-making, inclusive leadership and governance, and environmental policy frameworks that recognize Indigenous land rights and traditional ecological knowledge will help create biodiversity conservation models that are culturally sensitive and environmentally sustainable. Investments in capacity building, education, and technology transfer will further empower local populations, enabling them to take on leadership roles in biodiversity conservation and sustainable resource management.

As global environmental challenges such as man-made climate change, habitat loss, and zoonotic disease emergence continue to escalate, integrating holistic, community-driven approaches within *Conservation Medicine* will be critical. By bridging the gap between ecosystem preservation and socio-economic development, these integrated strategies can help build resilient ecosystems and thriving communities, ensuring a more sustainable future for both people and the planet.

5.7 References.

1. Aguirre, A. A., Ostfeld, R. S., Tabor, G. M., House, C., & Pearl, M. C. (2002). *Conservation medicine: Ecological health in practice*. Oxford University Press.
2. Anthony, S. J., Epstein, J. H., Murray, K. A., et al. (2013). A strategy to estimate unknown viral diversity in mammals. *mBio, 4*(5), e00598-13.
3. Banks-Leite, C., et al. (2014). Using ecological thresholds to evaluate the costs and benefits of set-asides in a biodiversity hotspot. *Science, 345*(6200), 1041-1045.
4. Berkes, F. (2018). *Sacred Ecology: Traditional Ecological Knowledge and Resource Management*. Routledge.

5. Berrahmouni, N., et al. (2016). The Great Green Wall: Implementation status and way ahead. *Food and Agriculture Organization (FAO)*.

6. Blehert, D. S., Hicks, A. C., Behr, M., et al. (2009). Bat white-nose syndrome: An emerging fungal pathogen? *Science, 323*(5911), 227.

7. Bonwitt, J., et al. (2018). Participatory epidemiology: Engaging communities in outbreak detection and response. *EcoHealth, 15*(3), 453-465.

8. Carroll, D., et al. (2018). Pathogen surveillance in conservation: A genomics-driven approach to spillover prevention. *Nature Communications, 9*(1), 1235.

9. Carroll, D., Daszak, P., Wolfe, N. D., et al. (2018). The global virome project. *Science, 359*(6378), 872-874.

10. Daszak, P., Cunningham, A. A., & Hyatt, A. D. (2000). Emerging infectious diseases of wildlife—Threats to biodiversity and human health. *Science, 287*(5452), 443-449.

11. Daszak, P., et al. (2020). Infectious disease emergence and the human-wildlife interface. *Science, 367*(6475), 379-382.

12. Díaz, S., et al. (2019). *IPBES Global Assessment Report on Biodiversity and Ecosystem Services*. Intergovernmental Science-Policy Platform on Biodiversity and Ecosystem Services (IPBES).

13. Dudley, N., et al. (2010). *Natural solutions: Protected areas helping people cope with climate change*. IUCN Report.

14. Frankham, R., et al. (2017). Genetic Management of Fragmented Animal and Plant Populations. *Oxford University Press*.

15. Garnett, S. T., et al. (2018). A spatial overview of the global importance of Indigenous lands for conservation. *Nature Sustainability, 1*(7), 369-374.

16. Gilbert, M., Golding, N., Zhou, H., et al. (2014). Predicting the risk of avian influenza A H7N9 infection in live-poultry markets across Asia. *Nature Communications, 5*, 4116.

17. Gilbert, M., et al. (2020). Canine distemper virus as a threat to wild carnivore conservation. *Viruses, 12*(11), 1324.

18. Groot, M. J., et al. (2020). One Health approach to wildlife conservation and zoonotic disease prevention. *Trends in Parasitology, 36*(2), 103-112.

19. Grooten, M., & Almond, R. E. A. (2018). *Living Planet Report 2018: Aiming higher.* World Wildlife Fund.

20. Hilty, J. A., et al. (2020). *Corridor Ecology: Linking Landscapes for Biodiversity Conservation and Climate Adaptation.* Island Press.

21. Jones, K. E., Patel, N. G., Levy, M. A., et al. (2008). Global trends in emerging infectious diseases. *Nature, 451*(7181), 990-993.

22. Jones, B. A., et al. (2013). Zoonosis emergence linked to agricultural intensification and environmental change. *Proceedings of the National Academy of Sciences, 110*(21), 8399-8404.

23. Karesh, W. B., Dobson, A., Lloyd-Smith, J. O., et al. (2012). Ecology of zoonoses: Natural and unnatural histories. *The Lancet, 380*(9857), 1936-1945.

24. Karesh, W. B., et al. (2012). Wildlife trade and global disease emergence. *Emerging Infectious Diseases, 18*(3), 202-207.

25. Keeley, A. T. H., et al. (2018). Connectivity can limit species' range shifts under climate change. *Proceedings of the National Academy of Sciences, 115*(49), 12263-12268.

26. Keesing, F., et al. (2010). Impacts of biodiversity on the emergence and transmission of infectious diseases. *Nature, 468*(7324), 647-652.

27. Knobel, D. L., et al. (2008). Re-evaluating the burden of rabies in Africa and Asia. *Bulletin of the World Health Organization, 86*(5), 360-368.

28. Kock, R. A. (2014). The role of wildlife in the epidemiology of Ebola virus disease (EVD). *Onderstepoort Journal of Veterinary Research, 81*(5), 8-13.

29. Kuiken, T., Leighton, F. A., Fouchier, R. A., et al. (2005). Pathogen surveillance in animals. *Science, 309*(5741), 1680-1681.

30. Laurance, W. F., et al. (2012). Averting biodiversity collapse in tropical forest protected areas. *Nature, 489*(7415), 290-294.

31. Mazet, J. A. K., Hall, J. S., Gill, J. S., Kock, R., & Clifford, D. L. (2016). A One Health" approach to address emerging zoonoses: The HALI Project in Tanzania. *PLOS Medicine, 13*(4), e1002003.

32. McDermott, J., et al. (2019). Sustainable development goals: Implications for global health. *The Lancet Global Health, 7*(6), e756-e757.

33. Murray, K. A., et al. (2022). One Health surveillance and pandemic preparedness: A systems approach. *The Lancet Planetary Health, 6*(2), e156-e165.

34. Naidoo, R., et al. (2016). Evaluating the effectiveness of community-based natural resource management: A case study from Namibia. *Biological Conservation, 200*, 143-152.

35. Noss, R. F., et al. (2012). From Yellowstone to Yukon: Conservation lessons from large-scale ecology. *Conservation Biology, 26*(5), 873-877.

36. Ostfeld, R. S., & Keesing, F. (2012). Effects of host diversity on infectious disease. *Annual Review of Ecology, Evolution, and Systematics, 43*(1), 157-182.

37. Patz, J. A., Epstein, P. R., Burke, T. A., & Balbus, J. M. (2005). Global climate change and emerging infectious diseases. *JAMA, 275*(3), 217-223.

38. Peres, C. A., et al. (2018). Biodiversity conservation in human-modified Amazonian forest landscapes. *Biological Conservation, 227*, 312-321.

39. Plowright, R. K., et al. (2017). Pathways to zoonotic spillover. *Nature Reviews Microbiology, 15*(8), 502-510.

40. Pretty, J., et al. (2009). The interplay between conservation and development: Case studies from around the world. *Environmental Conservation, 36*(3), 169-177.

41. Ralls, K., & Ballou, J. D. (2020). Captive breeding and reintroduction. *Conservation Biology for All, 2*(1), 364-380.

42. Randall, D. A., et al. (2020). Rabies vaccination of Ethiopian wolves using oral baiting. *Vaccine, 38*(4), 811-819.

43. Schroth, G., & McNeely, J. A. (2011). *Agroforestry and Biodiversity Conservation in Tropical Landscapes*. Island Press.

44. Schwab, S., et al. (2021). Advancements in oral wildlife vaccines: A new frontier for conservation medicine. *Current Opinion in Virology, 50*, 50-58.

45. Spenceley, A., et al. (2017). Sustainable tourism and conservation in Rwanda: Balancing biodiversity and development. *Journal of Sustainable Tourism, 25*(10), 1529-1546.

46. Stanley Price, M. R. (2018). Reintroduction of the Arabian oryx and Oryx leucoryx into Oman: Lessons learned for future reintroductions. *Oryx, 52*(4), 593-600.

47. Watson, J. E. M., et al. (2014). The performance and potential of protected areas. *Nature, 515*(7525), 67-73.

48. Zedler, J. B., & Kercher, S. (2005). Wetland resources: Status, trends, ecosystem services, and restorability. *Annual Review of Environment and Resources, 30*(1), 39-74.

6.0

Policy, Leadership, and Governance in Conservation Medicine.

The intricate challenges at the nexus of human, animal, and environmental health demand visionary leadership, cohesive governance, and evidence-based policymaking. This section delves into the frameworks, strategies, and transformational leadership approaches that shape effective *Conservation Medicine* initiatives. By exploring the evidence, policy innovations, and global collaborations, we highlight the indispensable role of governance in developing resilience and addressing interconnected health crises. Together, these insights illuminate a path forward toward sustainable and equitable solutions in *Conservation Medicine*.

6.1 Introduction.

Conservation Medicine confronts emerging health challenges stemming from ecosystem shifts, biodiversity decline, and human-induced pressures (Aguirre et al., 2012). Advancing this field requires robust policy, leadership, and governance, which determine how scientific insights are transformed into practical strategies that promote health, well-being, and resilience alongside environmental stewardship. This section explores the intricate interplay among these three pillars, providing insights into how decision-making structures, organizational leadership, and governance mechanisms influence the trajectory of conservation initiatives.

Policy in *Conservation Medicine* involves the development and implementation of regulations, guidelines, and international agreements that govern ecosystem health and biodiversity conservation protection (Karesh et al., 2012). Public health policies that incorporate *One Health* and *EcoHealth* approaches—emphasizing the inextricable interconnection between humans, animals, and ecosystems—serve as critical tools for mitigating zoonotic disease spillovers, antimicrobial resistance, and environmental degradation (Destoumieux-Garzón et al., 2018). Strong governance structures, such as global treaties and national conservation programs, ensure compliance with policies aimed at safeguarding *Planetary Health*.

Leadership in *Conservation Medicine* extends beyond governmental entities to include scientists, healthcare professionals, indigenous communities, and non-governmental organizations (NGOs) advocating for sustainable practices (Osofsky et al., 2005). Transformational leadership within this domain requires a vision that integrates ecosystem resilience, ethical considerations, and stakeholder collaboration to drive systemic change. Leaders play a crucial role in bridging the gap between scientific research and policy implementation by fostering cross-sectoral partnerships and promoting evidence-based decision-making.

Governance mechanisms within *Conservation Medicine* encompass the institutional structures, legislative frameworks, and inclusive processes that guide and regulate conservation efforts (Sutherland et al., 2011). Transparent and inclusive governance ensures equitable access to limited resources and strengthens global cooperation in addressing transboundary environmental threats such as deforestation, man-made climate change, and emerging infectious diseases (EIDs). By examining case studies of successful governance models, this section highlights best practices and challenges in implementing biodiversity conservation policies across diverse socio-political contexts.

The following discourse will provide an in-depth analysis of policy frameworks, leadership strategies, and governance models shaping conservation medicine. By integrating scientific, legal, and ethical perspectives, this discourse aims to offer a comprehensive understanding of how interdisciplinary and multisectoral collaboration can drive meaningful change in protecting both human and environmental health.

6.2 The Role of International Organizations in Conservation Medicine.

6.2.1 Introduction.

International organizations serve as cornerstone entities in advancing *Conservation Medicine* by promoting global cooperation, crafting regulatory frameworks, and guiding the implementation of evidence-based public health policies. These efforts are instrumental in addressing the complex interplay between human, animal, and environmental health. Organizations such as the World Health Organization (WHO), World Organisation for Animal Health (WOAH, formerly OIE), United Nations Environment Programme (UNEP), and International Union for Conservation of Nature (IUCN) play multifaceted roles in shaping the field. Through educational programing, essential knowledge is disseminated to empower practitioners and communities. Their research initiatives generate critical insights into health and conservation challenges, while their advocacy for policy reforms drives sustainable practices and informed value-driven decision-making. Additionally, these organizations invest in capacity-building endeavors, equipping stakeholders with the tools and expertise necessary to foster resilience and innovation in *Conservation Medicine*. Collectively, their contributions form the backbone of global efforts to protect biodiversity conservation, prevent zoonotic diseases, and ensure the health of interconnected systems.

6.2.2 World Health Organization (WHO): Bridging Public Health and Conservation.

The World Health Organization (WHO) plays an indispensable role in advancing *Conservation Medicine* by tackling the intricate public health challenges stemming from environmental degradation, ecosystem disturbances, biodiversity loss, and zoonotic diseases (WHO, 2022). At the heart of its efforts lies the *One Health* framework, which underscores the interdependence of human, animal, and environmental health. Through this paradigm, the WHO collaborates with global partners to address pressing threats such as emerging infectious diseases (EIDs), antimicrobial resistance, and health risks exacerbated by man-made climate change.

The WHO leverages its expertise to develop and implement strategies that integrate ecosystem and environmental health considerations into the broader framework of global health security. By fostering multi-sectoral collaborations and driving evidence-based initiatives, the WHO ensures that conservation efforts are closely aligned with global public health priorities. These efforts translate into public policies that not only safeguard biodiversity conservation but also enhance human health, resilience, and well-being in the face of an increasingly interconnected array of health challenges. In doing so, the WHO exemplifies a holistic approach to health governance, setting a global standard for linking *Conservation Medicine* with sustainable public health outcomes.

6.2.3 World Organisation for Animal Health (WOAH): Safeguarding Animal Health and Biodiversity Conservation.

The World Organisation for Animal Health (WOAH) plays a pivotal role in safeguarding animal health, a cornerstone of *Conservation Medicine*, by establishing and enforcing international standards for disease surveillance and monitoring, rapid response mechanisms, biosecurity protocols, and wildlife health management practices

(WOAH, 2021). Its comprehensive *Wildlife Health Framework* serves as a critical tool for monitoring diseases within wild animal populations, aiming to mitigate the risks of zoonotic spillovers while supporting efforts to conserve biodiversity conservation.

The WOAH also fosters collaboration with global veterinary services and other wildlife welfare stakeholders to ensure that animal health is seamlessly integrated into broader conservation strategies. Through this partnership-driven approach, the organization develops and promotes public health policies that align animal health management with ecosystem preservation and sustainability goals. By addressing the health challenges at the interface of wildlife, domestic animals, and human populations, WOAH exemplifies the interconnected principles of *Conservation Medicine* and contributes significantly to global health and environmental resilience.

6.2.4 United Nations Environment Programme (UNEP): Addressing Environmental Determinants of Health.

As the foremost environmental authority within the United Nations system, the United Nations Environment Programme (UNEP) holds a pivotal role in bridging ecosystem health with the well-being of humans and animals. UNEP's efforts are instrumental in addressing the interconnected challenges posed by environmental degradation, biodiversity loss, and public health risks. Through initiatives like the *UNEP-WHO-OIE Tripartite Alliance*, UNEP fosters interdisciplinary and multisectoral collaboration to combat deforestation, pollution, and habitat destruction—key drivers of biodiversity loss and emerging infectious diseases (EIDs). These initiatives exemplify UNEP's commitment to integrating conservation priorities with global health strategies, ensuring holistic approaches to sustainability.

Additionally, UNEP's flagship *Global Environmental Outlook Reports* serve as a powerful resource in illuminating the far-reaching consequences of environmental degradation on health and biodiversity. These reports guide international public health policy responses, shaping strategies that mitigate ecosystem damage while promoting resilience and sustainable development. By championing the integration of ecosystem health into public health policies and conservation frameworks, UNEP exemplifies the leadership needed to address the multifaceted challenges of the modern age. Its work not only highlights the urgency of global action but also sets a standard for collaborative and evidence-driven approaches to *Conservation Medicine* and *Planetary Health* (UNEP, 2021).

6.2.5 International Union for Conservation of Nature (IUCN): Biodiversity Conservation and Policy Development.

The International Union for Conservation of Nature (IUCN) plays a vital role in advancing *Conservation Medicine* by addressing species extinction risks, fostering ecosystem-based health interventions, and championing policies that protect biodiversity. Through its renowned initiative, the IUCN *Red List of Threatened Species*, the organization provides a scientifically grounded tool for monitoring wildlife populations and identifying conservation priorities. This globally recognized resource not only assesses the health of species but also informs targeted actions to mitigate extinction threats and preserve ecosystem balance (IUCN, 2020).

Beyond species assessment, the IUCN actively collaborates with governments, non-governmental organizations (NGOs), and other stakeholders to integrate health considerations into the management of protected areas. By embedding environmental, ecosystem, and public health goals within conservation strategies, the IUCN emphasizes the interconnectedness of biodiversity conservation and human well-being. Its efforts also extend to capacity-building initiatives, public policy advocacy, and the development of innovative frameworks that align global biodiversity conservation ambitions with practical, local implementation. As a thought leader in *Conservation Medicine*, the IUCN exemplifies the integration of science, governance, and community engagement in addressing critical challenges at the intersection of health and biodiversity conservation.

6.2.6 Conclusion.

International organizations serve as key architects in shaping *Conservation Medicine* by developing empirically-driven evidence-based global health standards and regulations, facilitating cross-sectoral collaboration, and driving public policy initiatives that acknowledge the intricate interdependence of human, animal, and environmental health. Through coordinated efforts, these organizations create frameworks that guide nations in addressing biodiversity loss, zoonotic disease emergence, and ecosystem degradation. By setting regulatory benchmarks and promoting evidence-based best practices, they ensure that *Conservation Medicine* remains a priority in global health security and environmental stewardship.

Beyond policy advocacy, international organizations play a critical role in funding research, supporting capacity-building initiatives, and fostering data-sharing networks that enhance global emergency preparedness, resilience, and response mechanisms. Their influence extends to shaping legislation, guiding sustainable land-use policies, and encouraging the implementation of *One Health* and *EcoHealth* approaches that integrate empirically-driven real science with real-world conservation efforts.

Through collaborative partnerships, these organizations also strengthen disease surveillance systems, improve wildlife health monitoring, and develop early warning mechanisms for emerging infectious diseases (EIDs). Their work reinforces the need for a unified, interdisciplinary and multisectoral approach to tackling health and environmental challenges, ensuring that *Conservation Medicine* remains a cornerstone of global efforts to protect both planetary and public health.

6.3 Public Health Policies for Integrated Systems of Health and Conservation Medicine.

6.3.1 Introduction.

Public health policies are instrumental in advancing Integrated Systems of Health (ISH) that align seamlessly with the principles and practices of *Conservation Medicine*. These policies draw on frameworks such as *One Health* and *EcoHealth*, which emphasize the profound inextricable interconnectedness between human, animal, and environmental health. By adopting these holistic ISH approaches, public health policymakers can address the shared determinants of health and the complex interdependencies that define our ecosystems.

Embedding conservation principles into public health strategies strengthens efforts to prevent diseases, particularly those emerging from zoonotic spillovers, while simultaneously promoting environmental stewardship and sustainability. Such public health policies help mitigate the impacts of habitat destruction, man-made climate change, and biodiversity loss, which are all drivers of public health challenges. Furthermore, they contribute to ecosystem resilience by fostering practices that protect and restore natural habitats, ensuring the sustainability of ecosystem services essential to human health, well-being, and resilience.

By prioritizing interdisciplinary and multisectoral collaboration and empirically-driven evidence-based approaches, these public health policies provide a comprehensive framework for addressing global health challenges. Their integration into governance systems can promote equitable access to resources, bolster climate adaptation strategies, and align conservation goals with sustainable development efforts, thereby advancing both *Planetary Health* and human resilience in a rapidly changing world.

6.3.2 One Health and EcoHealth Approaches in Public Health Policies.

The *One Health* approach, promoted by the World Health Organization (WHO), World Organisation for Animal Health (WOAH), United Nations Environment Programme (UNEP), and the Food and Agriculture Organization (FAO), advocates for multisectoral collaboration to address zoonotic diseases, antimicrobial resistance, and environmental health risks (WHO, 2022). Countries integrating the *One Health* framework into their public health policies have improved pandemic emergency preparedness, resilience, and response, wildlife disease surveillance and monitoring, and food safety standards (Destoumieux-Garzón et al., 2018). Similarly, the *EcoHealth* framework emphasizes ecosystem-based interventions, such as habitat restoration and sustainable land management, to mitigate health risks associated with environmental degradation (Wilcox & Ellis, 2006).

6.3.3 Foundational Public Health Policies Supporting Conservation Medicine.

1. *Zoonotic Disease Prevention and Surveillance Policies.*
   Many nations have adopted policies requiring integrated disease surveillance, monitoring, and rapid response that includes human, livestock, and wildlife health data. The *Global Early Warning System for Animal Diseases* (GLEWS+), a collaboration between WHO, FAO, and WOAH, exemplifies a public health policy-driven resourcefulness aimed at preventing zoonotic outbreaks by strengthening cross-sectoral monitoring and rapid response (FAO, 2021).

2. *Antimicrobial Resistance (AMR) Control Strategies.*
   Public health policies addressing AMR integrate *Conservation Medicine* principles and practices by promoting responsible

antibiotic use in both human healthcare and animal husbandry. The *WHO Global Action Plan on AMR* calls for reducing antibiotic overuse, enhancing infection prevention, and investing in alternative treatments that do not harm biodiversity conservation (WHO, 2021).

3. *Environmental Protection and Climate Change Policies.*
Recognizing that man-made climate change accelerates health risks, policies such as the *Paris Agreement (2015)* and the *UNEP-led Climate and Health Initiative* advocate for reducing GHG emissions and protecting natural habitats to minimize climate-related disease burdens (UNEP, 2021). These policies support *Conservation Medicine* by reinforcing the role of biodiversity conservation in regulating disease vectors and promoting ecosystem health.

4. *Food Safety and Sustainable Agriculture Policies.*
Integrated food safety policies incorporate Conservation Medicine principles and practices by enforcing sustainable agricultural practices, biodiversity conservation protection, and habitat conservation. The *Codex Alimentarius,* developed by FAO and WHO, establishes guidelines for food production that minimize environmental harm while ensuring public health safety and environmental stewardship (FAO/WHO, 2020).

5. *Wildlife and Ecosystem Protection Laws.*
Many nations design and implement conservation policies with direct public health implications, such as the *Convention on Biological Diversity (CBD)* and the *CITES Treaty (Convention on International Trade in Endangered Species of Wild Fauna and Flora)*. These public health policies aim to halt illegal wildlife trade, a significant driver of zoonotic disease spillovers (IUCN, 2020).

6.3.4 Conclusion.

Public health policies that incorporate Conservation Medicine principles and practices serve as a foundation for building sustainable, resilient healthcare delivery systems capable of rapidly responding to emerging global challenges. By integrating *One Health* and *EcoHealth* strategies, these public health policies promote a holistic ISH and comprehensive approach that recognizes the interdependence of human, animal, and environmental health. Governments and international organizations play a pivotal role in developing and implementing policies that prevent zoonotic disease outbreaks, combat antimicrobial resistance, protect biodiversity conservation, and mitigate the health impacts of man-made climate change.

Preventing zoonotic diseases requires robust surveillance systems, interdisciplinary and multisectoral collaboration, and regulatory frameworks that limit human-wildlife interactions, particularly in areas of rapid land-use change and biodiversity loss. Policies supporting integrated disease surveillance, such as the *Global Early Warning System for Animal Diseases (GLEWS+)*, enhance early detection and rapid response mechanisms, reducing the likelihood of pandemics.

Combating antimicrobial resistance (AMR) demands public health policies that regulate antibiotic use in human medicine, veterinary care, and agriculture while promoting research into alternative treatments. The *WHO Global Action Plan on AMR* serves as a model for harmonized international efforts aimed at reducing antibiotic overuse, improving stewardship programs, and investing in novel therapies.

Biodiversity conservation protection policies, including those established under the *Convention on Biological Diversity (CBD)* and the *CITES Treaty*, support Conservation Medicine by preserving ecosystems that provide essential disease-regulating services. Healthy ecosystems act as buffers against zoonotic disease transmission, and conservation-driven public health policies prioritize habitat preservation, reforestation efforts, and sustainable land management practices.

Climate-related health risks, such as vector-borne diseases and heat-related illnesses, are increasingly shaping global public health policy. Initiatives like the *UNEP Climate and Health Action Plan* highlight the need for climate adaptation strategies that integrate conservation efforts with integrated systems of health (ISH). Public health policies that promote carbon sequestration, sustainable food systems, and pollution reduction directly benefit both public health and environmental stewardship and stability.

Strengthening these policies will be critical for advancing global health security, ecosystem resilience, and environmental stewardship in an era of unprecedented ecosystem and health challenges. By embedding conservation medicine principles into public health governance and integrated systems of health (ISH), nations can create adaptive, forward-thinking ISHs that address the root causes of emerging infectious diseases (EIDs) while safeguarding biodiversity conservation and Planetary Health for future generations.

6.4 The Convention on Biological Diversity, One Health, and EcoHealth Collaborations.

6.4.1 Introduction.

The *Convention on Biological Diversity* (CBD), together with the *One Health* and *EcoHealth* frameworks, serves as a cornerstone in uniting biodiversity conservation with comprehensive public health strategies and innovations. These multisectoral and interdisciplinary collaborative approaches emphasize the profound and inextricable interconnections between human, animal, and environmental health, forming a cohesive foundation for addressing global biodiversity conservation challenges. By fostering environmental stewardship and advocating for the sustainable use and preservation of biodiversity, these frameworks underscore the essential role ecosystems play in supporting health and resilience across all facets of life.

The CBD's strategic objectives align with the principles and practices of *One Health* and *EcoHealth* frameworks by addressing the root causes of biodiversity loss and fostering public health policies that protect ecosystem integrity. These efforts are essential for mitigating the emergence and spread of infectious diseases, as the degradation of ecosystems often facilitates zoonotic spillovers. Furthermore, the *One Health* and *EcoHealth* frameworks prioritize the enhancement of ecosystem resilience, acknowledging that thriving ecosystems provide vital services such as climate regulation, water purification, and food security—all of which are indispensable for human and animal well-being.

By bridging conservation goals with public health priorities, the CBD and these interdisciplinary frameworks contribute to sustainable development and environmental stewardship on a global scale. They support creative and innovative approaches to governance, encourage collaborative research, and advocate for inclusive public health policies that balance ecosystem stewardship and sustainability with the health needs of diverse populations. Collectively, their integrated perspective underscores the importance of biodiversity conservation as a foundation for a healthier, more sustainable, and resilient future.

6.4.2 The Convention on Biological Diversity (CBD): A Global Framework for Conservation and Health.

The *CBD*, established in 1992 at the Earth Summit in Rio de Janeiro, was an international treaty aimed at promoting biodiversity conservation, sustainable use of natural resources, and fair sharing of genetic resources (CBD, 2022). The *Kunming-Montreal Global Biodiversity Framework*, adopted at CBD COP15 in 2022, set ambitious targets to halt biodiversity loss and restore ecosystems by 2030, emphasizing the role of healthy ecosystems in preventing zoonotic disease spillovers (CBD, 2022).

The *CBD's Health and Biodiversity Initiative*, in collaboration with the World Health Organization (WHO), advances research and public health policy integration on the links between biodiversity and health. This includes:

1. Strengthening *disease surveillance, monitoring, and rapid response* in biodiversity hotspots.
2. Promoting *nature-based solutions* for mitigation of man-made climate change and ecosystem and environmental health adaptation.
3. Supporting *traditional and indigenous knowledge* in biodiversity conservation and ecosystem service's management (WHO & CBD, 2021).

6.4.3 One Health: Bridging Human, Animal, and Environmental Health.

One Health is best known as a collaborative, multisectoral framework that addresses the interdependence of human, animal, and environmental health. Spearheaded by the World Health Organization (WHO), the World Organisation for Animal Health (WOAH), the Food and Agriculture Organization (FAO), and the United Nations Environment Programme (UNEP), *One Health* initiatives aim to prevent zoonotic disease spillover and outbreaks, combat antimicrobial resistance (AMR), and promote sustainable agricultural practices (WHO, 2022).

The *One Health Joint Plan of Action (2022-2026)* outlines the framework's key priorities:

1. Enhancing *disease surveillance and early warning systems* for rapidly responding to the earliest clinical signs of zoonotic and vector-borne diseases.
2. Strengthening *biosafety and biosecurity* in human-wildlife interactions.
3. Reducing *environmental drivers* of zoonotic and vector-borne diseases, such as deforestation and illegal wildlife trade (WHO, 2022).

6.4.4 EcoHealth: A Systems Approach to Health and Ecosystem Sustainability.

EcoHealth extends the *One Health* concept by emphasizing ecosystem-based interventions to improve health outcomes. *EcoHealth* research and education highlights the role of biodiversity conservation in regulating disease transmission and promotes public health policies that integrate ecosystem restoration with public health strategies and interventions (Wilcox & Ellis, 2006).

Key *EcoHealth* initiatives include:

1. *Reforestation and habitat restoration* to reduce zoonotic spillovers (Myers et al., 2013).
2. *Wetland conservation* to improve water quality and prevent vector-borne diseases (Patz et al., 2004).
3. *Community-led conservation programs* that integrate traditional ecosystem and environmental health knowledge with scientific research (Daszak et al., 2000).

6.4.5 CBD, One Health, and EcoHealth: A Collaborative Path Forward.

The intersection of *CBD, One Health,* and *EcoHealth* reflects a growing recognition that biodiversity conservation is critical for global health security. Future collaborations should focus on:

1. *Strengthening public health policy coherence* between environmental and healthcare sectors.
2. *Scaling up nature-based solutions (NbS)* for proactive health including disease prevention, health protection, health promotion, disease surveillance and population health management, and emergency preparedness, resilience, and response.
3. *Increasing funding for interdisciplinary research* on biodiversity conservation and health linkages.

6.4.6 Conclusion.

By embedding *Conservation Medicine* principles and practices into global public health policy frameworks, the *Convention on Biological Diversity (CBD), One Health,* and *EcoHealth* collaborations create a comprehensive foundation for safeguarding human, animal, and environmental health. These initiatives recognize that biodiversity conservation is not only an ecosystem priority but also a fundamental component of global health security, sustainable development, environmental stewardship, and climate change resilience.

The *CBD's Global Biodiversity Framework* establishes legally binding commitments from participating countries for ecosystem restoration, wildlife animal protection, and sustainable resource management, ensuring that biodiversity conservation efforts align with public health policy goals and objectives. Meanwhile, *One Health* initiatives drive interdisciplinary and multisectoral collaboration between human medicine, veterinary science, and environmental research, promoting public health policies that prevent zoonotic disease spillovers, curb antimicrobial resistance (AMR), and enhance global pandemic preparedness, resilience, and response. Simultaneously, *EcoHealth* principles emphasize ecosystem nature-based solutions—such as reforestation, wetland conservation, and sustainable agriculture—to mitigate health risks arising from environmental degradation.

Integrating these approaches into national and international public health policies strengthens disease surveillance systems, early warning systems, regulatory frameworks, and transboundary conservation efforts, fostering resilient societies capable of adapting to emerging infectious diseases (EIDs), advancing ecosystem service's management and overall health and well-being challenges. Investments in nature-based solutions (NbS), public health infrastructure, and cross-sectoral research and education further reinforces the long-term sustainability of integrated systems of health (ISH) and *Conservation Medicine.*

As the global community faces increasing threats from man-made climate change, habitat destruction, and infectious disease emergence, continued collaboration among the *CBD, One Health,* and *EcoHealth* networks will be essential in shaping a future where biodiversity conservation and the public's health are mutually reinforcing goals. By prioritizing empirically-driven, holistic, and evidence-based strategies, these frameworks pave the way for a healthier planet and a more resilient global society.

6.5 Leadership and Governance in Conservation Medicine.

6.5.1 Introduction.

Effective leadership and governance are foundational to advancing *Conservation Medicine* and tackling global health challenges arising from biodiversity loss, emerging infectious diseases (EIDs), and environmental degradation. Leadership in this domain demands a multidisciplinary and cross-sectoral approach, combining the expertise of public health professionals, human and veterinary medicine practitioners, ecologists, policymakers, and other stakeholders. This integration facilitates the development of sustainable, empirically-driven, and evidence-based solutions that address the interconnected health of humans, animals, and ecosystems.

Governance mechanisms play an equally pivotal role, encompassing regulatory frameworks, institutional collaborations, and policy enforcement strategies that ensure the principles of *Conservation Medicine* are operationalized effectively. These mechanisms establish the infrastructure needed to promote transparency, accountability, and coordinated action at national and international scales. By fostering global partnerships, aligning policy with scientific evidence, and implementing standardized practices, governance enables *Conservation Medicine* initiatives to achieve measurable impact, mitigate health risks, and enhance ecosystem resilience.

Together, effective leadership and robust governance create the foundation for transformational approaches in *Conservation Medicine*, promoting equity, sustainability, and resilience in the face of pressing environmental and health challenges.

6.5.2 Leadership in Conservation Medicine.

Leadership in *Conservation Medicine* extends beyond traditional governmental and scientific institutions. It involves global organizations, policymakers, local communities, Indigenous leaders, and non-governmental organizations (NGOs) working collectively to promote biodiversity conservation and health security.

1. *Interdisciplinary and Collaborative Leadership.*
   Conservation Medicine requires leaders who can bridge disciplines such as epidemiology, wildlife management, and environmental science. The *One Health* and *EcoHealth* frameworks have emphasized the need for interdisciplinary leadership, fostering cooperation between the World Health Organization (WHO), the World Organisation for Animal Health (WOAH), the Food and Agriculture Organization (FAO), and the United Nations Environment Programme (UNEP) (WHO, 2022).

2. *Scientific Leadership and Research Innovation.*
   Scientific leadership drives innovation in disease surveillance, ecosystem health monitoring, and climate adaptation strategies. Institutions such as the *EcoHealth Alliance* and the *International Union for Conservation of Nature (IUCN)* provide critical research insights, influencing public health policy recommendations and global conservation priorities (Daszak et al., 2000).

3. *Community and Indigenous Leadership.*
   Local and Indigenous communities play a crucial role in Conservation Medicine, contributing traditional ecological

knowledge (TEK) that enhances biodiversity conservation management. Research and education highlight the role of Indigenous land stewardship in protecting ecosystems and preventing zoonotic disease spillovers (Garnett et al., 2018). Leadership models that integrate community-based conservation approaches have demonstrated long-term success in sustainable health and environmental outcomes.

6.5.3 Governance in Conservation Medicine.

Governance in *Conservation Medicine* includes the legal, institutional, and public health policy structures that regulate human interactions with the environment and wildlife. Effective governance requires strong regulatory frameworks, enforcement mechanisms, and international cooperation to address biodiversity conservation and health challenges.

1. *International Treaties and Agreements.*
   a) *The Convention on Biological Diversity (CBD)* provides a legal framework for protecting biodiversity conservation while recognizing its direct impact on human health (CBD, 2022).
   b) *The Convention on International Trade in Endangered Species (CITES)* regulates the trade of wildlife species to prevent biodiversity loss and disease transmission (CITES, 2021).
   c) *The Paris Agreement* addresses Anthropocene climate change-related health risks, advocating for public health policies that integrate *Conservation Medicine* principles and practices into climate adaptation strategies (UNEP, 2021).

2. *Public Health Policy Development and Enforcement.*
   National governments worldwide are pivotal in advancing public health and environmental policies that align with the principles and practices of *Conservation Medicine*. One notable example is the *Global Virome Project*, an ambitious initiative supported by multinational governments and international environmen-

tal stewardship organizations. This project aims to proactively identify and mitigate zoonotic disease threats by mapping viruses with the potential to spill over into human populations, thereby addressing risks before they can emerge (Carroll et al., 2018). Through coordinated efforts, this initiative exemplifies the critical role of global collaboration in safeguarding health and biodiversity conservation.

3. *Ethical and Equitable Governance.*
   Governance models must ensure that *Conservation Medicine* policies are inclusive, ethical, and equitable. Transparency in decision-making, community participation, and fair resource allocation are essential to addressing health disparities and environmental justice concerns in conservation efforts (Sutherland et al., 2011).

6.5.4 Conclusion.

Leadership and governance in *Conservation Medicine* are fundamental to advancing interdisciplinary and multisectoral collaboration, driving public health policy innovation, and building sustainable integrated systems of health (ISH) that are instrumental in achieving human, animal, and environmental well-being. As global challenges such as biodiversity loss, man-made climate change, emerging infectious diseases (EIDs), and antimicrobial resistance (AMR) escalate, strong leadership and effective governance frameworks are required to ensure a coordinated and proactive response.

Successful *Conservation Medicine* governance depends on the ability to bridge diverse sectors, including public health, human and veterinary medicine, ecology, and policy-making. By combining scientific research, education, Indigenous knowledge, and evidence-based regulatory frameworks, leaders and policymakers can develop resilient and adaptive governance systems that safeguard biodiversity conservation while bolstering global health security.

Traditional Ecological Knowledge (TEK), an invaluable resource preserved by Indigenous and local communities, offers profound insights into environmental and ecosystem service's management, and health prevention, protection, and promotion. Incorporating *TEK* into governance strategies enriches decision-making processes, ensuring they are inclusive, sustainable, and culturally competent and empathetic to the complexities of ecosystems and community resilience. This integration not only strengthens biodiversity conservation efforts but also fosters equitable and collaborative approaches to addressing interconnected global challenges.

Strengthening international cooperation and local engagement will be critical in shaping a future where biodiversity conservation and positive health outcomes are mutually reinforcing priorities. Global frameworks such as the *Convention on Biological Diversity (CBD), One Health,* and *EcoHealth* provide models for integrated governance, promoting transboundary collaboration to mitigate disease risks, protect natural habitats, and regulate wildlife trade. Additionally, fostering community-led conservation initiatives, cross-sectoral policymaking, and public-private partnerships will enhance the long-term effectiveness of *Conservation Medicine* efforts.

By prioritizing holistic, empirically-driven, evidence-based, and equity-focused public health policies, leaders can build governance structures that are responsive, adaptable, and capable of addressing both current and emerging threats to *Planetary Health*. This approach will ensure that *Conservation Medicine* remains a cornerstone of global efforts to achieve sustainability, resilience, and biodiversity conservation protection for future generations.

6.6 References.

1. Aguirre, A. A., Ostfeld, R. S., Tabor, G. M., House, C., & Pearl, M. C. (Eds.). (2012). *Conservation medicine: Ecological health in practice*. Oxford University Press.

2. Carroll, D., Daszak, P., Wolfe, N. D., et al. (2018). The Global Virome Project. *Science, 359*(6378), 872-874.

3. Convention on International Trade in Endangered Species of Wild Fauna and Flora (CITES). (2021). *CITES and One Health: Managing Wildlife Trade for a Sustainable Future.*

4. Convention on Biological Diversity (CBD). (2022). *Kunming-Montreal Global Biodiversity Framework.*

5. Destoumieux-Garzón, D., Mavingui, P., Boetsch, G., et al. (2018). The One Health concept: 10 years old and a long road ahead. *Frontiers in Veterinary Science, 5*, 14.

6. Daszak, P., Cunningham, A. A., & Hyatt, A. D. (2000). Emerging infectious diseases of wildlife–Threats to biodiversity and human health. *Science, 287*(5452), 443-449.

7. Food and Agriculture Organization (FAO). (2021). *GLEWS+: The Global Early Warning System for Animal Diseases.*

8. Food and Agriculture Organization (FAO) & World Health Organization (WHO). (2020). *Codex Alimentarius: International Food Standards.*

9. Garnett, S. T., Burgess, N. D., Fa, J. E., et al. (2018). A spatial overview of the global importance of Indigenous lands for conservation. *Nature Sustainability, 1*(7), 369-374.

10. International Union for Conservation of Nature (IUCN). (2020). *The IUCN Red List of Threatened Species.*

11. Karesh, W. B., Dobson, A., Lloyd-Smith, J. O., et al. (2012). Ecology of zoonoses: Natural and unnatural histories. *The Lancet, 380*(9857), 1936-1945.

12. Myers, S. S., Gaffikin, L., Golden, C. D., et al. (2013). Human health impacts of ecosystem alteration. *Proceedings of the National Academy of Sciences, 110*(47), 18753-18760.

13. Osofsky, S. A., Kock, R. A., Kock, M. D., Kalema-Zikusoka, G., & Grahn, R. (2005). Building support for protected areas using a "One Health" perspective. *Conservation Biology, 19*(5), 1343-1348.

14. Patz, J. A., Daszak, P., Tabor, G. M., et al. (2004). Unhealthy landscapes: Policy recommendations on land use change and infectious disease emergence. *Environmental Health Perspectives, 112*(10), 1092-1098.

15. Sutherland, W. J., Dicks, L. V., Everard, M., & Geneletti, D. (2011). The value of conservation scientists engaging in policy. *Conservation Biology, 25*(6), 1185-1191.

16. United Nations Environment Programme (UNEP). (2021). *Preventing the Next Pandemic: Zoonotic Diseases and How to Break the Chain of Transmission.*

17. United Nations Environment Programme (UNEP). (2021). *Climate, Environment, and Health Action Plan.*

18. Wilcox, B. A., & Ellis, B. (2006). Forests and emerging infectious diseases of humans. *UNESCO-SCOPE Policy Briefs, 13*, 1-12.

19. World Health Organization (WHO). (2021). *Global Action Plan on Antimicrobial Resistance.*

20. World Organisation for Animal Health (WOAH). (2021). *Wildlife Health Framework.*

21. World Health Organization (WHO) & Convention on Biological Diversity (CBD). (2021). *Biodiversity and Health Linkages: An Integrated Approach to Public Health and Conservation.*

22. World Health Organization (WHO). (2022). *One Health Joint Plan of Action (2022-2026).*

7.0

Vital Directions in Conservation Medicine.

The future of *Conservation Medicine* will be shaped by the increasing recognition of the intricate relationships between human health, wildlife health, and the environment. As global challenges such as man-made climate change, habitat destruction, and emerging infectious diseases continue to escalate, *Conservation Medicine* must evolve with innovative strategies to address these pressing concerns (Daszak et al., 2001).

One of the critical areas for future development in *Conservation Medicine* is the integration of artificial intelligence (AI) and big data analytics. Advances in technology now allow for more precise monitoring of disease outbreaks in wildlife populations and predictive modeling of zoonotic spillover events (Carroll et al., 2018). AI-driven disease surveillance and rapid response systems can help researchers detect early warning signs of pandemics and assess ecological disruptions before they reach critical levels (Chowdhury et al., 2021).

Another vital direction involves expanding *One Health* and *EcoHealth* initiatives that emphasize collaboration among veterinarians, physicians, ecologists, and public health professionals. Strengthening interdisciplinary frameworks will improve global responses to zoonotic threats while fostering sustainable solutions for biodiversity conservation (Destoumieux-Garzón et al., 2018).

This collaborative approach will enhance disease surveillance efforts, promote responsible land use policies, and mitigate human-wildlife conflict (Wolfe et al., 2007).

Climate change adaptation strategies will also play a significant role in shaping the future direction of *Conservation Medicine*. As global temperatures rise and extreme weather events become more frequent, species migration patterns, ecosystem stability, and vector-borne disease distributions will shift unpredictably. *Conservation Medicine* must prioritize climate resilience research, develop adaptive management plans, and advocate for policies that protect vulnerable species and habitats (Hoberg & Brooks, 2015).

In addition to technological and interdisciplinary advancements, ethical considerations must guide the future of *Conservation Medicine*. Balancing conservation goals with community needs requires culturally sensitive approaches that engage and empower local populations in sustainable conservation practices (Redford et al., 2011). By incorporating indigenous knowledge and community-based participatory research methods, conservation efforts can be more equitable and effective.

Ultimately, the future of *Conservation Medicine* depends on proactive global policies, cross-sectoral collaboration, and scientific innovation. With a growing understanding of the inextricable interconnectedness between human, animal, and ecosystem health, *Conservation Medicine* is well-positioned to address 21st-century ecosystem and public health challenges.

7.1 Strengthening Global Surveillance and Early Warning Systems.

7.1.1 Introduction.

Conservation Medicine is fundamentally anchored in the ability to monitor and respond quickly to emerging infectious diseases (EIDs) on a global scale. Robust disease surveillance and effective early warning systems are crucial tools in identifying potential threats at their inception, enabling timely intervention to protect both ecosystem and human health. These systems act as the first line of defense against the emergence and spread of zoonotic diseases—pathogens that can jump from animals to humans—by tracking patterns of disease transmission within wildlife populations, livestock, and human communities.

Strengthening these surveillance mechanisms is imperative to reducing the risk of zoonotic spillover events, where infectious agents cross the species barrier. Such events can have devastating consequences, including the loss of biodiversity, the disruption of delicate ecosystems, and significant public health crises. By detecting unusual disease patterns early, *Conservation Medicine* practitioners can work collaboratively across disciplines and borders to implement mitigation strategies, such as improved biosecurity measures, vaccination campaigns, and habitat restoration efforts.

Moreover, these systems provide critical data to inform policy decisions, facilitate interdisciplinary research, and enhance global preparedness for future pandemics. A well-functioning surveillance network not only protects human populations but also helps safeguard the resilience of ecosystems, ensuring that the interconnected health of humans, animals, and the environment is preserved for generations to come. Investing in these efforts is a cornerstone of promoting Planetary Health and addressing the complex challenges posed by our rapidly changing world.

7.1.2 The Need for Enhanced Disease Surveillance.

Emerging infectious diseases (EIDs) are increasingly linked to wildlife, with over 70% of human EIDs originating as zoonotic infections—diseases transmitted between animals and humans (Jones et al., 2008). These zoonotic diseases often emerge at the interface of human and animal ecosystems, highlighting the profound interconnectedness of biodiversity conservation and public health. Anthropogenic activities, such as man-made climate change, habitat destruction, and global trade, significantly amplify the risk of zoonotic spillover events, creating conditions that enable pathogens to cross species barriers with greater frequency and intensity (Daszak et al., 2020).

Anthropogenic climate change drives ecosystem disruptions, forcing wildlife to migrate into new territories, often closer to human settlements, in search of food and suitable habitats. Simultaneously, habitat destruction—resulting from deforestation, agricultural expansion, and urbanization—brings humans, domestic animals, and wildlife into closer contact. These interactions increase the probability of pathogen spillover, as they disrupt natural barriers that typically prevent pathogens in wildlife populations from infecting humans. Additionally, global trade and travel create pathways for pathogens to spread rapidly across regions and continents, amplifying the risk of local outbreaks escalating into global health crises.

In this context, the need for a well-integrated and proactive disease surveillance system cannot be overstated. Such a system serves as an indispensable tool for detecting early signals of disease outbreaks, enabling timely intervention to contain and mitigate their impacts. Surveillance networks must operate across multiple scales—local, national, and international—and integrate large amounts of data and information from diverse sources, including wildlife monitoring, environmental health assessments, and public health records.

By leveraging real-time data and advanced scientific technologies, such as metagenomics and artificial intelligence, these systems can identify potential threats before they escalate into widespread pandemics. Furthermore, an effective surveillance framework requires interdisciplinary and multisectoral collaboration among ecologists, veterinarians, physicians, public health professionals, and policymakers to ensure comprehensive monitoring, rapid response, and the development of proactive measures.

Investing in robust disease surveillance infrastructure not only protects human health but also safeguards biodiversity conservation and ecosystem stability. As the drivers of EIDs become increasingly global and complex, strengthening these systems is essential to preserving the delicate balance between human and ecosystem health. This integrated and comprehensive approach embodies the principles and practices of *One Health*, emphasizing the interconnectedness of humans, animals, and ecosystems in combating the growing threat of emerging infectious diseases.

7.1.3 Key Strategies for Strengthening Global Surveillance and Early Warning Systems.

1. *One Health Approach* – Implementing a *One Health* framework enhances data sharing across human, animal, and ecosystems sectors, improving outbreak prediction and response (Mackenzie & Jeggo, 2019).
2. *Genomic and AI-Based Monitoring* – Advances in metagenomics and artificial intelligence (AI) enhance pathogen detection in wildlife reservoirs, identifying novel threats before human spillover (Kupferschmidt, 2021).
3. *Community Engagement and Citizen Science* – Involving local communities in wildlife monitoring enhances data collection in remote regions, improving disease intelligence (Watsa et al., 2020).
4. *Global Data Integration* – Strengthening data-sharing platforms such as WHO's Epidemic Intelligence from Open Sources (EIOS) and the Global Virome Project (Carroll et al., 2018) ensures a coordinated international response.
5. *Predictive Modeling and Risk Mapping* – Leveraging ecosystem and epidemiological modeling aids in identifying high-risk areas for zoonotic emergence (Allen et al., 2017).

7.1.4 Policy and Funding Imperatives.

Effective disease surveillance hinges on sustained investment in global health security to ensure preparedness, response, and resilience against emerging infectious diseases (EIDs). The interconnected nature of global healthcare delivery systems demands coordinated efforts that transcend national borders, as no single country can address the complex and dynamic risks of zoonotic diseases in isolation. Strengthening early warning systems is a foundational element of such efforts, allowing for the timely detection of unusual patterns in disease transmission and enabling rapid, targeted interventions to contain outbreaks before they escalate.

International collaboration is paramount in building and maintaining these robust systems. Organizations such as the World Health Organization (WHO), the World Organisation for Animal Health (WOAH), and the Coalition for Epidemic Preparedness Innovations (CEPI) play pivotal roles in this global network. These entities facilitate the exchange of critical data, establish standardized protocols for disease reporting, and provide technical guidance to harmonize efforts across countries. Their collaborative frameworks help to integrate human, animal, and ecosystem health perspectives, embodying the *One Health* approach to global health security.

However, to maximize the effectiveness of these initiatives, financial resources must be strategically allocated to address the most vulnerable regions. Funding must prioritize capacity-building in low-resource settings, particularly in areas where the risk of zoonotic disease emergence is highest. This includes supporting the development of laboratory and public health infrastructure, training, recruitment, and retention of healthcare and veterinary professionals, and fostering community engagement and empowerment to improve early detection and response mechanisms.

By equipping these regions with the tools and knowledge needed to identify and respond to potential infectious disease outbreaks, the global health community can significantly reduce the risk of localized events escalating into global health emergencies. Furthermore, targeted investments in research, education, and innovation—such as genomic sequencing, data-sharing platforms, and artificial intelligence-driven analytics—can enhance the efficiency and accuracy of surveillance systems, ensuring they remain adaptable to evolving threats.

In the face of increasing pressures from man-made climate change, habitat loss, and globalized trade, sustained and equitable investments in global health security are more urgent than ever. A collective and well-funded approach not only strengthens disease surveillance but also protects human lives, preserves biodiversity conservation, and fosters resilience within ecosystems, ultimately safeguarding *Planetary Health*.

7.1.5 Conclusion.

Global surveillance and early warning systems in *Conservation Medicine* must continuously evolve to effectively address the increasing complexity and frequency of modern disease threats. The rapid expansion of human populations, climate change, habitat destruction, and globalization have intensified interactions between humans, animals, and ecosystems, creating greater opportunities for zoonotic spillover and biodiversity-related health crises.

To enhance the ability to predict, detect, and mitigate emerging infectious diseases (EIDSs), an interdisciplinary and multisectoral approach is necessary. Integrating advanced medical and scientific technologies such as genomic surveillance, artificial intelligence-driven predictive modeling, and real-time data analytics will significantly improve pathogen detection and risk assessment. These tools enable early identification of novel threats, allowing for timely intervention before outbreaks escalate.

Fostering international collaboration is equally critical, as infectious diseases do not recognize borders. Strengthening global partnerships among governments, research institutions, public health agencies, and conservation organizations will facilitate data sharing, capacity building, and coordinated outbreak responses. Initiatives like the World Health Organization's *Epidemic Intelligence from Open Sources (EIOS)* and the *Global Virome Project* exemplify the power of collaborative efforts in disease surveillance, monitoring and response.

Moreover, embracing a *One Health* approach—an interdisciplinary framework that integrates human, animal, and ecosystem health—will provide a more comprehensive understanding of disease dynamics. By prioritizing cross-sectoral cooperation and investing in sustainable ecosystem management, the root causes of emerging infectious diseases can be mitigated sooner rather than solely responding to outbreaks after they occur.

Ultimately, modernizing global surveillance and early warning systems requires sustained investment, public and environmental policy innovation, and a commitment to proactive health security. Strengthening these systems will not only mitigate the risks of future pandemics but also safeguard biodiversity conservation and promote global health resilience.

7.2 Integrating technological advancements in Conservation Medicine.

7.2.1 Introduction.

Conservation Medicine is increasingly harnessing cutting-edge scientific and medical technologies to address complex challenges at the intersection of ecosystem and human health. These advancements are transforming traditional approaches to disease surveillance, biodiversity conservation, and ecosystem management, enabling more precise, efficient, and predictive solutions.

Emerging technologies such as artificial intelligence (AI) are revolutionizing data analysis and decision-making. AI-driven tools can process vast datasets to identify patterns, predict disease outbreaks, and optimize conservation strategies. Machine learning algorithms, for instance, can model the spread of zoonotic pathogens, aiding in the development of targeted interventions to prevent spillover events. Additionally, AI facilitates real-time monitoring of ecosystems by analyzing data collected from sensors, drones, and other remote sensing technologies.

Genomic sequencing is another powerful innovation, providing insights into the genetic makeup of pathogens and wildlife populations. This technology allows researchers to trace the origins of emerging infectious diseases, monitor genetic diversity in threatened species, and identify potential vulnerabilities in ecosystems. Tools like environmental DNA (eDNA) analysis are further expanding the scope of genomic technologies by enabling non-invasive monitoring of biodiversity in aquatic and terrestrial environments.

Remote sensing technologies, including satellite imagery and drone-based data collection, are transforming the way scientists observe and manage ecosystems. These tools provide high-resolution data on habitat changes, deforestation rates, and wildlife migration patterns, offering a comprehensive view of ecosystem dynamics. Such technologies are particularly valuable in inaccessible or fragile environments, where on-the-ground monitoring may be limited.

Big data analytics complements these advancements by integrating diverse datasets—from genomic sequences to ecosystem monitoring records—into actionable insights. This holistic and integrated approach enables scientists to detect early signals of ecosystem disruptions, predict trends, and develop adaptive management strategies. For example, predictive models powered by big data can anticipate the impacts of Anthropogenic climate change on species distribution, guiding conservation efforts toward the most vulnerable regions.

By leveraging these emerging advanced scientific and medical technologies, *Conservation Medicine* is becoming increasingly proactive in its efforts to safeguard human, animal, and ecosystem health. The integration of AI, genomics, remote sensing, and big data analytics fosters interdisciplinary and multisectoral collaboration and empowers scientists to address global threats with unprecedented precision, scalability, and spread. These innovations are paving the way for a more resilient and sustainable future, where biodiversity conservation and public health are deeply intertwined.

7.2.2 Artificial Intelligence and Machine Learning.

AI and machine learning (ML) have become powerful tools in *Conservation Medicine*, enabling real-time disease monitoring, predictive modeling, and automated data analysis. AI-driven platforms analyze vast datasets from diverse sources—including satellite imagery, social media reports, and genomic data—to identify emerging infectious disease's hotspots and forecast potential outbreaks (Murray et al., 2022). For instance, AI-powered tools have been instrumental in tracking zoonotic disease spillover risks by analyzing ecosystem and epidemiological trends (Ostfeld et al., 2021).

7.2.3 Genomic Surveillance and Metagenomics.

Advancements in genomic sequencing have improved the detection and characterization of pathogens circulating in wildlife and ecosystem reservoirs. Metagenomic approaches allow scientists to analyze microbial communities in water, soil, and animal populations, identifying novel pathogens before they pose a threat to human or animal health (Carroll et al., 2018). Programs such as the *Global Virome Project* utilize high-throughput genomic sequencing to catalog potential zoonotic viruses, enhancing early warning capabilities (Anthony et al., 2017).

7.2.4 Remote Sensing and Geographic Information Systems (GIS).

Satellite-based remote sensing and GIS technologies provide critical insights into ecosystem changes that influence disease dynamics. These tools help monitor deforestation, climate variability, and urban expansion—factors that contribute to the displacement of wildlife and increased human-animal interactions (Gibb et al., 2020). By mapping high-risk areas for zoonotic spillover, researchers and policymakers can implement targeted conservation strategies to mitigate disease emergence.

7.2.5 Big Data and Predictive Analytics.

The integration of big data analytics allows for the synthesis of vast amounts of ecosystem, epidemiological, and genomic data, enabling more precise risk assessments. Predictive analytics models, which incorporate climate, land-use, and pathogen prevalence data, can help forecast future outbreaks and inform early intervention strategies (Allen et al., 2017). Cloud-based platforms facilitate real-time data sharing among global health and conservation organizations, fostering collaborative responses to emerging threats (Daszak et al., 2020).

7.2.6 Conclusion.

Scientific and medical technological advancements are revolutionizing *Conservation Medicine* by significantly improving disease surveillance, ecosystem monitoring, and predictive modeling. As human activities continue to impact ecosystems, the risk of zoonotic spillover, biodiversity loss, and ecosystem services disruptions increases. Cutting-edge tools such as artificial intelligence (AI), genomic sequencing, remote sensing, and big data analytics are enhancing our ability to detect, predict, and mitigate emerging health threats at the human-animal-ecosystem interface. These innovations allow for real-time tracking of disease outbreaks, identification of high-risk areas for pathogen emergence, and proactive responses to ecosystem changes that strongly influence the public's health.

The integration of AI-driven platforms enables automated data processing from diverse sources, such as satellite imagery, wildlife monitoring, and global health databases, improving the accuracy and speed of outbreak predictions. Genomic surveillance provides critical insights into the evolutionary trajectories of pathogens, facilitating early detection of novel diseases in wildlife populations before they spread to humans. Meanwhile, remote sensing technologies and Geographic Information Systems (GIS) offer valuable tools for monitoring deforestation, habitat fragmentation, and climate-driven changes that contribute to emerging infectious diseases (EIDs). By synthesizing these technological advancements with big data analytics, *Conservation Medicine* is becoming more proactive rather than reactive, allowing for targeted interventions that safeguard both ecosystem and human health.

However, to fully leverage these technological advancements, continued investment in research, education, infrastructure, and global collaboration is imperative. Strengthening interdisciplinary partnerships between ecologists, epidemiologists, data scientists, and policymakers will ensure the effective implementation of these technologies. Additionally, fostering international cooperation and data-sharing initiatives will enhance our collective ability to prevent and mitigate health crises before they escalate. By embracing these innovations and reinforcing cross-sectoral collaboration, *Conservation Medicine* can play a pivotal role in building a more resilient global healthcare delivery system while protecting biodiversity conservation and ecosystem integrity.

7.3 Expanding *One Health* and *EcoHealth* Approaches in Conservation Medicine.

7.3.1 Introduction.

Conservation Medicine is increasingly adopting interdisciplinary frameworks such as *One Health* and *EcoHealth* to address the complex interactions between human, animal, and ecosystem health. These approaches recognize that human health, well-being, and resilience is fundamentally linked to ecosystem integrity and biodiversity conservation. As global ecosystem changes intensify, expanding these frameworks is critical to preventing emerging infectious diseases (EIDs), mitigating biodiversity loss, and fostering sustainable health systems.

7.3.2 The *One Health* Approach in Conservation Medicine.

The *One Health* approach champions cross-disciplinary collaboration between human medicine, veterinary medicine, and environmental sciences to develop holistic, integrated, and sustainable health solutions (Mackenzie & Jeggo, 2019). By recognizing the inextricable interconnectedness of human, animal, and environmental health, this model is essential to *Conservation Medicine*, especially in addressing zoonotic diseases—infectious diseases that originate in animals and account for over 70% of emerging infectious diseases (EIDs) globally (Jones et al., 2008). These infectious diseases are often the result of human activities that disrupt ecosystems, such as deforestation, Anthropocene climate change, and land-use changes, underscoring the need for integrated frameworks like *One Health*.

Applying *One Health* principles and practices to conservation efforts brings several key benefits. It facilitates early detection of infectious disease threats through comprehensive early warning disease surveillance systems that monitor wildlife, livestock, and human populations. By identifying potential zoonotic spillover risks early, stakeholders can implement targeted interventions to mitigate outbreaks. Additionally, *One Health* supports responsible land-use planning, helping to preserve natural habitats and reduce human-wildlife interactions that heighten spillover risks. The approach also promotes biodiversity conservation as a cornerstone of health resilience, recognizing that healthy ecosystems provide vital barriers against pathogen emergence.

Real-world applications of *One Health* have demonstrated its transformational potential in combating infectious disease threats, such as Ebola, avian influenza, and COVID-19. For instance, Uganda's *PREDICT program*, part of a global zoonotic disease surveillance initiative, has proven the effectiveness of *One Health* in identifying high-risk spillover zones, strengthening interdisciplinary and multisectoral collaboration, and enhancing preparedness, monitoring and rapid response key performance indicators (Mazet et al., 2020). By integrating data across disciplines, the program has provided actionable insights into infectious disease dynamics, enabling timely interventions that protect both human health and ecosystems.

Scaling up and spreading *One Health* principles and practices in *Conservation Medicine* is essential to addressing the growing threat of zoonotic diseases and ecosystem disruptions. Key strategies include expanding early waring disease surveillance systems to monitor emerging infectious disease threats globally, investing in interdisciplinary research that bridges fields like genomics and epidemiology, and incorporating *One Health* principles and practices into education and professional training programs. Furthermore, fostering international cooperation is critical, as global health challenges require collective action and knowledge sharing among nations, institutions, and communities.

By embedding *One Health* into *Conservation Medicine* practices, the global community can enhance its ability to detect and prevent emerging infectious diseases' health threats, preserve biodiversity conservation, and ensure the sustainability of ecosystems for future generations. This holistic and integrated systems approach is not just a scientific necessity—it is a moral imperative to address the interconnected health challenges of an increasingly complex world.

7.3.3 The *EcoHealth* Approach: A Systems Perspective in Conservation Medicine.

The *EcoHealth* framework prioritizes ecosystem-based solutions to health challenges by integrating social, ecological, and biomedical sciences to address the fundamental drivers of infectious disease emergence (Wilcox & Ellis, 2006). This holistic and integrated systems approach emphasizes the interconnectedness of human, animal, and environmental health while addressing the ecosystem and societal factors that contribute to the spread of infectious diseases. Unlike the *One Health* framework, which predominantly focuses on zoonotic disease transmission and its intersections with human and veterinary medicine, *EcoHealth* adopts a broader systems perspective. It considers upstream drivers such as deforestation, agricultural intensification, man-made climate change, and socioeconomic factors that accelerate infectious disease emergence and transmission (Morand & Lajaunie, 2021).

Through this expansive lens, *EcoHealth* has provided valuable insights into the impact of land-use changes on the ecology of vector-borne diseases like malaria and Lyme disease. For example, deforestation and habitat fragmentation often disrupt the natural balance between vectors, such as mosquitoes and ticks, and their host species. These disruptions can lead to an increase in vector populations or force wildlife into closer proximity to human settlements, raising the likelihood of pathogen spillover. Habitat alteration also reshapes the distribution and behavior of both vectors and pathogens, creating new dynamics in disease transmission that heighten risks to public health (Gibb et al., 2020).

The *EcoHealth* approach offers actionable pathways for addressing these challenges by promoting sustainable land management practices. These include preserving and restoring natural habitats to maintain biodiversity conservation, which acts as a protective buffer against infectious disease transmission. For example, biodiverse ecosystems often dilute the transmission of pathogens, as diverse host species reduce the probability of vectors encountering susceptible hosts. Similarly, reforestation and habitat connectivity initiatives help to stabilize ecosystems, mitigating the pressures that lead to spillover events.

Moreover, integrating *EcoHealth* principles practices into *Conservation Medicine* can enhance efforts to safeguard both public health and biodiversity conservation. Restoring degraded ecosystems, implementing agroecological farming practices, and prioritizing environmental stewardship are not merely conservation strategies—they are public health strategies. They tackle the root causes of infectious disease emergence by addressing the ecosystems' services disruptions that facilitate pathogen transmission. *EcoHealth* also encourages the involvement of local communities in conservation and health initiatives, recognizing the importance of Indigenous knowledge and participatory governance in achieving sustainable positive health outcomes.

By combining ecosystem sustainability with health priorities, the *EcoHealth* framework provides a comprehensive strategy to reduce the risks of infectious disease emergence, protect biodiversity conservation, and promote ecosystem resilience. It underscores the necessity of an integrated, systems-thinking approach to addressing the global health challenges of the 21st century, where environmental and public health are inseparably linked.

7.3.4 Integrating *One Health* and *EcoHealth* for Global Health Security.

Expanding *One Health* and *EcoHealth* approaches in Conservation Medicine requires an integrated strategy that combines early warning disease surveillance and response, ecosystem protection, and community engagement. Key priorities include:

1. *Enhancing Global Surveillance* – Strengthening disease monitoring networks such as the *Global Virome Project* and WHO's *Epidemic Intelligence from Open Sources (EIOS)* to detect emerging threats early (Carroll et al., 2018).
2. *Investing in Sustainable Development* – Promoting upstream public health policies that reduce deforestation, regulate wildlife trade, and support ecosystem services restoration to minimize zoonotic spillover risks (Daszak et al., 2020).
3. *Fostering Interdisciplinary Collaboration* – Encouraging partnerships between public health agencies, conservation biologists, veterinarians, and public health policymakers to develop holistic and comprehensive interventions (Ostfeld et al., 2021).
4. *Community-Based Approaches* – Engaging local populations in disease surveillance, wildlife conservation, and ecosystem management to improve resilience against health threats (Watsa et al., 2020).

7.3.5 Conclusion.

Expanding the *One Health* and *EcoHealth* approaches in *Conservation Medicine* has the potential to fundamentally reshape global health security by addressing the interconnected challenges of emerging infectious diseases, biodiversity loss, and environmental degradation. These interdisciplinary frameworks offer a proactive, systems-based strategy that integrates human, animal, and ecosystem health to prevent and mitigate global health risks.

By strengthening these approaches, *Conservation Medicine* can move beyond reactive responses to outbreaks and instead focus on early detection, risk reduction, and sustainable management of ecosystems that influence disease dynamics. A robust *One Health* approach facilitates cross-sectoral collaboration between public health experts, veterinarians, ecologists, and policymakers, ensuring a more coordinated response to zoonotic threats. Simultaneously, an *EcoHealth* perspective promotes sustainable land-use practices, ecosystem restoration, and climate resilience—key factors in reducing human-wildlife conflict and limiting pathogen spillover.

The integration of these two frameworks serves as a vital blueprint for preventing future pandemics, protecting biodiversity conservation, and strengthening the resilience of both human populations and ecosystems' services amidst global change. Investing in global surveillance systems, community engagement, and interdisciplinary research, education, and practice will be critical in operationalizing these approaches at multiple socio-ecological levels of organization. By prioritizing the health of ecosystems alongside human and animal health, *Conservation Medicine* can play a transformational role in building a more resilient, sustainable, and health-secure future for generations to come.

7.4 Funding Mechanisms for Conservation Medicine.

7.4.1 Introduction.

Conservation Medicine requires sustainable funding to support research, education, surveillance, practice, and intervention programs. Traditional commercial and governmental health insurance for humans may cover infectious disease diagnostics and treatment, but animal and ecosystem health require long-term public and private funding. As global ecosystem changes accelerate the emergence of zoonotic diseases and biodiversity loss, securing robust financial support is critical to advancing *Conservation Medicine* initiatives. Funding mechanisms for *Conservation Medicine* come from various sources, including government agencies, international organizations, private foundations, and innovative financial instruments.

7.4.2 Government and Multilateral Funding.

Governments play a central role in financing *Conservation Medicine* through public health, environmental health, and agricultural agencies. National and international programs such as the U.S. Centers for Disease Control and Prevention (CDC), United States Agency for International Development (USAID), and European Union Horizon Europe provide substantial funding for research, education, and implementation of *One Health* initiatives (Karesh et al., 2012). Multilateral organizations, including the World Health Organization (WHO) and the World Bank, allocate limited resources world-wide for pandemic emergency preparedness and *Conservation Medicine* projects in high-risk regions (McElwee et al., 2020).

Examples of government-backed funding programs include:

1. *The Global Virome Project*, supported by USAID, which aims to catalog high-risk viruses in wildlife to prevent zoonotic spillover (Carroll et al., 2018).

2. *WHO's Pandemic Fund,* which provides financial support for early warning and disease surveillance systems that align with conservation medicine objectives (World Bank, 2022).
3. *The European Green Deal,* which funds projects at the inter-section of ecosystem conservation and public health resilience (European Commission, 2021).

7.4.3 Private Foundations and Philanthropy.

Private foundations and philanthropic organizations have become increasingly involved in *Conservation Medicine* funding. Major contributors include:

1. *The Bill & Melinda Gates Foundation,* which funds zoonotic infectious disease surveillance and *One Health* research (Gates Foundation, 2020).
2. *The Wellcome Trust,* which invests in antimicrobial resistance (AMR) and the impact of ecosystem service changes on disease emergence (Wellcome, 2021).
3. *The Rockefeller Foundation,* which has historically supported *EcoHealth* and other systemic sustainable health initiatives (Rockefeller Foundation, 2020).

These organizations often provide research grants, fellowships, and institutional funding to support interdisciplinary and multi-sectoral studies on emerging infectious diseases (EIDs), ecosystem health, and wildlife conservation.

7.4.4 Public-Private Partnerships (PPPs) and Corporate Funding.

Collaboration between public institutions and private entities has led to innovative financing models for *Conservation Medicine.* Public-private partnerships (PPPs) facilitate long-term investments in infectious disease surveillance, sustainable agriculture, and biodiversity conservation. Corporate funding, particularly from industries such as pharmaceuticals, biotechnology, and agribusiness, contributes to *Conservation Medicine* initiatives by funding research and technology-driven solutions (Sokolow et al., 2019).

For instance, pharmaceutical companies invest in *One Health* initiatives to develop vaccines for zoonotic diseases, while agribusinesses support projects that promote sustainable livestock farming and prevent cross-species disease transmission (Destoumieux-Garzón et al., 2018).

7.4.5 Innovative Financing Mechanisms.

New financial instruments are emerging to support conservation medicine, including:

1. *Green Bonds and Biodiversity Credits* – Investment tools that raise capital for projects linking ecosystem health and disease prevention (Waldron et al., 2020).
2. *Payments for Ecosystem Services (PES)* – Programs that compensate communities for conservation efforts that reduce disease risks, such as maintaining wetlands to control vector-borne diseases (Pattanayak & Wendland, 2007).
3. *Climate and Health Insurance Schemes* – Financial products that provide funding for conservation medicine interventions in response to environmental health crises (Bowles et al., 2022).

7.4.6 Crowdfunding and Citizen Science Support.

With the rise of digital platforms, crowdfunding campaigns and citizen-led fundraising efforts have become viable sources of funding for *Conservation Medicine*. Organizations such as *Wildlife Conservation Society (WCS)* and *EcoHealth Alliance* have successfully used crowdfunding to support field research, wildlife disease monitoring, and educational initiatives (Watsa et al., 2020).

7.4.7 Conclusion.

A diverse range of funding mechanisms, spanning from traditional government grants to cutting-edge financial instruments, is crucial for sustaining and advancing *Conservation Medicine* initiatives. As the field continues to grow in response to global challenges such as biodiversity loss, Anthropocene climate change, and emerging infectious diseases (EIDs), securing sustainable and scalable financial resources is imperative. Multisectoral collaboration, innovative financing, and strategic investments must be prioritized to ensure the long-term impact and sustainability of *Conservation Medicine* programs worldwide.

Expanding financial resources through public-private partnerships, philanthropic contributions, and market-based financing will enhance global capacity to implement *Conservation Medicine* strategies effectively. Government funding remains a cornerstone of support, providing essential resources for research, infectious disease surveillance, and public and environmental health policy implementation. However, private sector engagement—including investments from biotechnology, pharmaceutical, and agribusiness industries—can drive innovation, research, education, and technological advancements in infectious disease monitoring and ecosystem health management. Philanthropic organizations and non-governmental organizations (NGOs) play a critical role by funding conservation projects that integrate health and environmental stewardship and sustainability.

Beyond traditional funding streams, innovative financing mechanisms such as green bonds, biodiversity credits, payments for ecosystem services (PES), and impact investing offer promising opportunities to generate sustained financial support. These market-driven approaches align economic incentives with conservation goals, encouraging sustainable land-use practices and ecosystem services' preservation efforts that reduce infectious disease risks and promote global health security. Additionally, crowdsourced funding and citizen science initiatives are emerging as viable sources of micro-financing, engaging and empowering communities to participate in *Conservation Medicine* efforts.

Strengthening and diversifying financial mechanisms for *Conservation Medicine* is essential to building resilience against current and future health and ecosystem crises. By leveraging interdisciplinary and multisectoral partnerships, sustainable financial models, and forward-thinking investment strategies, *Conservation Medicine* can continue to safeguard both human and ecosystem health. A well-financed, globally coordinated approach will not only mitigate future pandemics and environmental degradation but also foster sustainable ecosystem services' stability and global health security.

7.5 Workforce Development in Conservation Medicine.

7.5.1 Introduction.

As *Conservation Medicine* grows as a critical field at the nexus of human, animal, and ecosystem health, the development of a highly skilled, interdisciplinary, and multisectoral workforce is paramount. This field addresses complex, interconnected challenges, such as emerging infectious diseases, biodiversity loss, and ecosystem services' degradation, which require expertise that spans multiple disciplines and sectors.

Professionals in *Conservation Medicine* must possess diverse clinical and non-clinical skill sets that include epidemiology, veterinary medicine, human medicine, ecology, public health, socioeconomics, and public and environmental policy-making. These disciplines provide the foundational knowledge and tools necessary to investigate, address, and mitigate the root causes of existential health threats (e.g., Anthropocene climate change with GHG emissions, global rise in pandemics and natural disasters) at the interface of ecosystems and societies. Beyond technical expertise, practitioners must also excel in communication, systems thinking, and stakeholder engagement to foster collaboration among multidisciplinary teams and communities.

Workforce development in *Conservation Medicine* should emphasize four key pillars:

1. *Rigorous Real Science Education*: Training programs must integrate real-world scientific approaches, emphasizing critical thinking, evidence-based practices, logical reasoning, and interdisciplinary problem-solving. Universities and other academic institutions must collaborate to create specialized curricula that combine systems biology, healthcare systems science, and environmental studies, preparing professionals for the unique demands of this field.

2. *Skill-Building and Practical Training*: Professionals require hands-on training in areas such as disease surveillance, monitoring, and rapid response, environmental risk assessment, and public health policy. Internships, fieldwork, and research opportunities should be incorporated into educational pathways to enhance applied knowledge and foster creativity and innovation in addressing human, animal and ecosystem health challenges.

3. *Interdisciplinary and Multisectoral Collaboration*: *Conservation Medicine* thrives on the integration of knowledge from diverse sectors, including government agencies, non-governmental organizations (NGOs), and international bodies. Workforce development must encourage partnerships that break down communication silos, enabling professionals to work collaboratively across disciplines and cultural contexts.

4. *Capacity Strengthening in Resource-Limited Settings*: Addressing global health security demands a focus on low-to-middle income regions, where zoonotic disease emergence often occurs. Investments in specialized training and infrastructure must be tailored to these settings, enabling local communities to lead health and biodiversity conservation efforts. Capacity strength-

ening includes not only technical training but also empowering local leaders and leveraging Indigenous knowledge systems.

By fostering a workforce with interdisciplinary and multisectoral expertise and the ability to adapt to diverse environmental and societal contexts, *Conservation Medicine* delivers creative and innovative solutions to pressing global challenges. This approach ensures that practitioners are equipped to design and implement strategies that bridge health and environmental priorities, creating a sustainable future and resilience. Prioritizing equitable workforce development across both high-income and resource-limited regions is key to building a globally coordinated response to health and conservation needs.

7.5.2 Education and Training Programs.

Expanding educational opportunities in *Conservation Medicine* is critical to developing a skilled workforce. Universities, academic and research institutions worldwide have established specialized *One Health* and *EcoHealth programs*, which integrate conservation science with medical and veterinary training (Mazet et al., 2020). Graduate programs, such as those at the University of California, Davis and the London School of Hygiene & Tropical Medicine, emphasize interdisciplinary approaches that prepare students to address complex health and environmental challenges.

Additionally, professional training and certification programs are increasing in popularity. Organizations such as the *Wildlife Conservation Society (WCS)* and the *EcoHealth Alliance* offer training workshops on zoonotic disease surveillance, environmental and ecosystem health monitoring, and environmental risk assessment (Destoumieux-Garzón et al., 2018). These programs enhance practitioners' ability to detect and mitigate emerging infectious diseases linked to ecosystem services' disturbances.

7.5.3 Interdisciplinary and Cross-Sector Collaboration.

Building a workforce that can effectively operate across disciplines is crucial for the success of *Conservation Medicine*. Experts from public health, human and veterinary medicine, ecology, and environmental science must work collaboratively to develop holistic and integrated health solutions. Joint training programs, such as the *One Health Workforce* initiatives supported by the U.S. Agency for International Development (USAID), have successfully built multisectoral teams across Africa and Southeast Asia (Mackenzie & Jeggo, 2019).

Moreover, public-private partnerships between academia, government agencies, and industries such as pharmaceuticals and biotechnology contribute to knowledge sharing and practical applications in *Conservation Medicine* (Ostfeld et al., 2021). Encouraging professionals to engage in interdisciplinary and multisectoral collaborations strengthens the field's ability to respond to emerging existential health threats efficiently.

7.5.4 Workforce Capacity Strengthening in Low-Resource Settings.

Given that many emerging infectious diseases (EIDs) originate in biodiversity hotspots, investing in workforce development in resource-limited regions (e.g., LMIC) is crucial (Morand & Lajaunie, 2021). Countries in the Global South, where zoonotic spillover risks are high, often face shortages of trained personnel and infrastructure to conduct early-warning disease surveillance and ecosystem services' monitoring. Initiatives such as the *PREDICT* project and FAO's *Emergency Prevention System (EMPRES)* have helped train local professionals in field epidemiology, wildlife disease tracking, and biosecurity measures (Carroll et al., 2018).

Community engagement and capacity-building efforts should be prioritized to empower local communities to contribute to *Conservation Medicine* efforts. Programs that involve indigenous knowledge and citizen science initiatives have proven effective in enhancing disease detection and ecosystem protection while fostering local leadership in conservation efforts (Watsa et al., 2020).

7.5.5 Integrating Technology in Workforce Training.

Scientific and medical technological advancements such as AI-driven disease modeling, remote sensing, and genomic surveillance are redesigning *Conservation Medicine*, necessitating new skill sets and training among professionals (Kupferschmidt, 2021). Digital learning platforms, virtual reality simulations, and real-time data-sharing networks provide innovative ways to train and connect *Conservation Medicine* experts globally. Investments in digital literacy and technological capacity-building will ensure that future professionals can harness these tools effectively.

7.5.6 Public and Environmental Policy and Funding for Workforce Development.

To sustain long-term workforce growth, governments, research institutes, public health agencies, and international relief organizations must prioritize public and environmental policies and funding mechanisms that support workforce development programs (Daszak et al., 2020). Increased investment in scholarships, research grants, and professional development opportunities will attract more individuals to the field and ensure that expertise is equitably distributed worldwide. Organizations such as the World Health Organization (WHO) and the World Bank have highlighted the importance of strengthening workforce capacity as a key pillar of global health security (World Bank, 2022).

7.5.7 Conclusion.

Building a robust, interdisciplinary workforce in *Conservation Medicine* is critical for addressing the complex and interconnected existential health threats that arise at the human-animal-ecosystem interface. This workforce must be equipped with the skills and knowledge necessary to confront emerging infectious diseases (EIDs), environmental degradation, and biodiversity loss. As these challenges intensify, the need for specially-trained professionals who can integrate public health, veterinary medicine, human medicine, environmental science, and epidemiology has never been greater.

Expanding education and training opportunities is the first step in cultivating the next generation of *Conservation Medicine* experts. Universities, research institutions, and professional organizations must develop comprehensive curricula that combine traditional disciplines with newer fields such as *EcoHealth*, *One Health*, wildlife disease management, and zoonotic disease surveillance, monitoring, and rapid response programs. The interdisciplinary and multisectoral programs should emphasize practical, hands-on training in diverse settings, from urban to suburban to rural ecosystems. Such training should be designed to meet the growing demand for professionals who can apply *One Health and EcoHealth* principles and practices, which recognize the inextricable link between the health of humans, animals, and ecosystems (Mackenzie & Jeggo, 2019).

Moreover, fostering interdisciplinary and multisectoral collaboration is essential to tackling global health challenges efficiently and effectively. *Conservation Medicine* is inherently interdisciplinary, requiring expertise from diverse sectors, including public health, human medicine, veterinary science, ecology, health systems science, and environmental and ecosystem management. Encouraging partnerships between academia, government agencies, NGOs, and the private sector can ensure that *Conservation Medicine* initiatives are both practical, scalable, and spreadable. International collaboration, particularly in areas most affected by emerging infectious diseases

(EIDs), such as biodiversity hotspots, is essential to share knowledge, pool workforce resources, and tackle shared risks (Ostfeld et al., 2021). By promoting interdisciplinary education and research, *Conservation Medicine* can strengthen the global health workforce and foster more effective responses to complex health and environmental threats.

In regions that face high risks from zoonotic diseases and environmental disruptions, such as tropical rainforests, wetlands, and biodiversity hotspots, investing in workforce capacity and retention is particularly urgent. Developing targeted specialized training programs in these areas, especially in low- and middle-income countries (LMIC), is essential for building local expertise. These regions often lack sufficient infrastructure and skilled professionals to respond to health crises. Thus, targeted investments in capacity-building initiatives—such as community health training, field epidemiology, community-based participatory research (CBPR), and wildlife monitoring—can empower local populations and strengthen the global response to emerging infectious diseases (Carroll et al., 2018). Additionally, these programs should incorporate local knowledge and cultures, as local communities play a key role in biodiversity conservation and disease prevention efforts (Watsa et al., 2020).

As scientific and medical technological advancements continue to revolutionize the field of *Conservation Medicine*, integrating advanced technology into training programs is crucial. The adoption of remote sensing tools, artificial intelligence (AI), and genomic surveillance can help professionals better understand disease dynamics, predict future health risks, and develop targeted interventions. Training programs should include digital literacy and equip professionals with the necessary tools to use cutting-edge technologies in real-world contexts. Telemedicine and online platforms can also enhance the accessibility of training for remote or underserved regions, providing experts with access to global knowledge networks. By embracing these technologies, *Conservation Medicine* can significantly increase the efficiency and reach of its initiatives (Kupferschmidt, 2021).

Finally, to ensure long-term sustainability, strategic public and environmental policy and financial investments must be sustained. Governments, international organizations, and the private sector should work together to prioritize funding for education, training, and workforce development in conservation medicine. This includes offering scholarships, grants, and financial incentives to support the growth of the field. Policies that promote interdisciplinary research, international collaboration, and technology integration are also essential to addressing the root causes of health and ecological challenges. Only through sustained investment in human capital can the *Conservation Medicine* workforce remain equipped to respond to emerging existential health threats, protect biodiversity conservation, and foster ecosystem resilience.

Building a robust, interdisciplinary, and multisectoral workforce in *Conservation Medicine* is essential to protecting global health and preserving the environment. By advancing comprehensive educational programs, fostering cross-sectoral partnerships, strengthening workforce capacity, and incorporating cutting-edge technological innovations into training, a resilient cadre of professionals can be equipped to tackle the intricate challenges at the intersection of human, animal, and ecosystem health. To ensure the continued growth and effectiveness of *Conservation Medicine*, strategic investments in public and environmental policy development and financial resources are critical. These investments will empower the field to play a pivotal role in preventing disease, enhancing ecosystem health, and conserving biodiversity, securing a healthier and more sustainable future for all.

7.6 References.

1. Allen, T., Murray, K. A., Zambrana-Torrelio, C., et al. (2017). Global hotspots and correlates of emerging zoonotic diseases. *Nature Communications, 8*(1), 1124.

2. Anthony, S. J., Gilardi, K., Menachery, V. D., et al. (2017). Further evidence for bats as the evolutionary source of Middle East respiratory syndrome coronavirus. *mBio, 8*(2), e00373-17.
3. Bowles, D. C., Butler, C. D., & Friel, S. (2022). Climate change, health, and economic benefits of action. *The Lancet Planetary Health, 6*(3), e183-e189.
4. Carroll, D., Daszak, P., Wolfe, N. D., et al. (2018). The Global Virome Project. *Science, 359*(6378), 872–874.
5. Chowdhury, N., Hossain, M. S., & Kashem, M. A. (2021). Artificial intelligence in wildlife conservation and disease monitoring: A systematic review. *Ecological Informatics*, 63, 101297.
6. Daszak, P., Cunningham, A. A., & Hyatt, A. D. (2001). Anthropogenic environmental change and the emergence of infectious diseases in wildlife. *Acta Tropica*, 78(2), 103-116.
7. Daszak, P., Olival, K. J., & Li, H. (2020). A strategy to prevent future pandemics: Integrating ecological and epidemiological approaches. *New England Journal of Medicine, 382*(21), 1993–1995.
8. Destoumieux-Garzón, D., Mavingui, P., Boetsch, G., et al. (2018). The One Health concept: 10 years old and a long road ahead. *Frontiers in Veterinary Science*, 5, 14.
9. European Commission. (2021). European Green Deal: Investing in a sustainable future. *EU Policy Briefs*.
10. Gates Foundation. (2020). The role of philanthropy in pandemic preparedness. *Gates Policy Reports*.
11. Gibb, R., Franklinos, L. H. V., Redding, D. W., & Jones, K. E. (2020). Ecosystem perspectives are essential for infectious disease ecology. *Nature Ecology & Evolution, 4*(11), 1471–1472.
12. Hoberg, E. P., & Brooks, D. R. (2015). Evolution in action: Climate change, biodiversity dynamics, and emerging infectious disease. *Philosophical Transactions of the Royal Society B: Biological Sciences*, 370(1665), 20130553.
13. Jones, K. E., Patel, N. G., Levy, M. A., et al. (2008). Global trends in emerging infectious diseases. *Nature, 451*(7181), 990–993.

14. Karesh, W. B., Dobson, A., Lloyd-Smith, J. O., et al. (2012). Ecology of zoonoses: Natural and unnatural histories. *The Lancet, 380*(9857), 1936–1945.

15. Kupferschmidt, K. (2021). How AI and genomic surveillance can prevent pandemics. *Science, 372*(6543), 778–781.

16. Mackenzie, J. S., & Jeggo, M. (2019). The One Health approach—Why is it so important? *Tropical Medicine and Infectious Disease, 4*(2), 88.

17. Mackenzie, J. S., & Jeggo, M. (2019). The One Health approach—Why is it so important? *Tropical Medicine and Infectious Disease, 4*(2), 88.

18. Mazet, J. A., Joly, D. O., & Kahn, L. H. (2020). One Health: From concept to practice. *Conservation Medicine, 2*(4), 12–18.

19. McElwee, P., Turnhout, E., Chiroleu-Assouline, M., et al. (2020). Ensuring a sustainable future: Conservation financing mechanisms. *Nature Sustainability, 3*(7), 555-563.

20. Morand, S., & Lajaunie, C. (2021). Biodiversity and health: Linking life, ecosystems, and societies. *Elsevier Science & Technology*.

21. Murray, K. A., Navarro, C., & Daszak, P. (2022). AI for pandemics: Leveraging artificial intelligence to predict, prevent, and respond to emerging infectious diseases. *Lancet Digital Health, 4*(7), e482-e490.

22. Ostfeld, R. S., Keesing, F., & Eviner, V. T. (2021). Infectious disease ecology: Effects of ecosystems on disease and of disease on ecosystems. *Princeton University Press*.

23. Pattanayak, S. K., & Wendland, K. J. (2007). Nature's care: Diarrhea, watershed protection, and biodiversity conservation in Flores, Indonesia. *Biodiversity and Conservation, 16*(10), 2801-2819.

24. Redford, K. H., Amato, G., Baillie, J., et al. (2011). What does it mean to successfully conserve a (vertebrate) species? *Bioscience*, 61(1), 39-48.

25. Rockefeller Foundation. (2020). The role of nature-based solutions in global health. *Rockefeller Reports on Health and Sustainability*.

26. Sokolow, S. H., Nova, N., Pepin, K. M., et al. (2019). Ecological interventions to prevent and manage zoonotic pathogen spillover. *Philosophical Transactions of the Royal Society B, 374*(1782), 20180342.

27. Waldron, A., Miller, D. C., Rusch, G. M., A. V., Adams, V., & Agarwal, B. (2020). Protecting 30% of the planet for nature: Costs, benefits, and economic implications. *Cambridge Conservation Initiative.*

28. Watsa, M., Pandey, S., Pérez-Álvarez, L., & Sanderson, C. E. (2020). Wildlife disease surveillance: A way forward. *Philosophical Transactions of the Royal Society B, 375*(1812), 20190589.

29. Wellcome Trust. (2021). Science to solve the health challenges of our time. *Wellcome Trust Annual Report.*

30. Wilcox, B. A., & Ellis, B. (2006). Forests and emerging infectious diseases of humans. *UNEP/WHO Report on Ecosystems and Health*, 1–28.

31. Wolfe, N. D., Dunavan, C. P., & Diamond, J. (2007). Origins of major human infectious diseases. *Nature*, 447(7142), 279-283.

32. Wood, J. L. N., Leach, M., Waldman, L., et al. (2021). A framework for the study of zoonotic disease emergence and its drivers: Spillover of bat coronaviruses as a case study. *Philosophical Transactions of the Royal Society B, 376*(1837), 20200359.

33. World Bank. (2022). Pandemic Fund: Strengthening global health security. *World Bank Health Reports.*

34. World Bank. (2022). Strengthening workforce capacity for global health security. *World Bank Health Reports.*

8.0

Conclusion.

Conservation Medicine is a dynamic, interdisciplinary and multisectoral field that delves into the profound interconnectedness of human, animal, and ecosystem health. Situated at the intersection of multiple scientific disciplines—epidemiology, ecology, veterinary medicine, human medicine, environmental science, and public health—it addresses mounting threats that disrupt the intricate balance of global biodiversity conservation. This integrative approach seeks to uncover and mitigate the impacts of ecosystem service degradation, human-driven climate change, and emerging infectious diseases (EIDs) that pose risks to both wildlife populations and human communities. As the global biodiversity crisis intensifies, *Conservation Medicine* stands as a critical tool in the mission to protect ecosystems, prevent species extinction, and shield humanity from the growing threats of zoonotic diseases.

The significance of *Conservation Medicine* has become especially evident in the wake of global health emergencies such as the COVID-19 pandemic. These crises have underscored the fragile relationship between human health, animal health, and the stability of ecosystem services. As zoonotic diseases arise with increasing frequency due to deforestation, habitat loss, and the encroachment of human activities into wildlife territories, the urgency for coordinated, cross-disciplinary solutions have never been greater.

In *Conservation Medicine: A Multidisciplinary Approach to Maintaining Global Biodiversity,* the integration of diverse fields of expertise is explored as a means to address the critical challenges at the human-animal-ecosystem interface. The book illustrates how *Conservation Medicine* can prevent, detect, and respond to existential health threats—ranging from pandemics to natural disasters—that transcend national borders. This makes it an indispensable component of global health security and ecosystem resilience. By uniting knowledge from human, veterinary, and environmental sciences, *Conservation Medicine* provides a robust, integrated and comprehensive system for tackling the root causes of EIDs and ecosystem degradation.

Key strategies discussed throughout the book include enhancing early-warning disease surveillance systems, strengthening wildlife health monitoring, and mitigating environmental risks through the *One Health* and *EcoHealth* frameworks. From pioneering technological innovations such as AI-powered early-warning disease surveillance to fostering multisectoral collaborations among governments, NGOs, and private stakeholders, the future of *Conservation Medicine* is rich with transformational potential. It offers a pathway toward creating a healthier, more sustainable planet.

However, realizing this vision requires sustained investment in workforce development, research, education, and global partnerships. The next critical steps involve scaling and spreading successful *Conservation Medicine* models to address urgent challenges such as biodiversity loss, habitat fragmentation, and human-induced climate change. As the field continues to evolve, it is essential that *Conservation Medicine* remains a central priority within global public health and environmental policy agendas. By ensuring that future generations inherit a planet where biodiversity thrives, human health is safeguarded, and ecosystems are resilient, we can achieve a balance where both humanity and nature coexist and flourish. Through collective action, collaboration, and unwavering commitment across disciplines and sectors, we can secure a brighter, more harmonious future for all life on Earth.

8.1 A Call to Action.

As the world teeters on the brink of an unprecedented environmental health crisis, the urgency of coordinated, comprehensive action has never been more apparent. The future of global biodiversity conservation—and by extension, the health, resilience, and well-being of humanity—hinges on the choices we make today. The challenges explored throughout this book underscore the intricate interconnections between human, animal, and ecosystem health. From the alarming rise of zoonotic diseases fueled by habitat destruction and anthropogenic climate change to the accelerating decline in biodiversity conservation, the call for a unified, global response is more critical than ever.

Conservation Medicine, with its multidisciplinary and integrative approach, holds immense potential for addressing these complex challenges. By bridging the gaps between environmental science, medicine, and public health policy, it offers a pathway to not only understanding but also mitigating these interconnected crises. To fully realize its transformational impact, the field demands robust support from governments, international organizations, private-sector partnerships, and local communities. Only through sustained investment, collaboration, and commitment can *Conservation Medicine* rise to meet the urgent and profound challenges facing the planet and pave the way for a more sustainable and resilient future.

A global call to action in *Conservation Medicine* must emphasize interdisciplinary collaboration, innovation, and the adoption of sustainable practices on a worldwide scale. Governments have a crucial role in driving this effort by investing in scientific research, developing early-warning surveillance systems, enhancing rapid response mechanisms, and strengthening workforce capacity to detect and address emerging health threats at the intersection of human, animal, and ecosystem health.

Global health organizations, including the World Health Organization (WHO) and the United Nations (UN), must lead the charge in advancing public and environmental policies that incorporate *One Health* and *EcoHealth* frameworks. These initiatives ensure that the interconnected health of ecosystems, wildlife, and human populations is considered equitably in environmental conservation strategies. Policies should aim to integrate diverse disciplines and promote collaborative, systems-based approaches to solving complex health and environmental challenges.

A comprehensive strategy must also include robust environmental monitoring systems designed to assess the functionality of ecosystem services and identify species at risk due to human activities, such as climate change, deforestation, and industrialization. These monitoring systems play a vital role in providing real-time data to inform evidence-based interventions, guiding efforts to mitigate the impacts of environmental degradation and safeguard biodiversity.

By aligning global policy and investment with the principles of *Conservation Medicine*, we can create a cohesive framework that supports the health of all living beings while fostering resilience in the face of global environmental changes. This unified approach offers an opportunity to not only address immediate challenges but also pave the way for a sustainable and equitable future.

Public-private partnerships play a critical role in leveraging resources and technological innovations to develop a more adaptive and sustainable *Conservation Medicine* framework. By fostering collaboration across sectors, these partnerships can accelerate the integration of advanced technologies, such as artificial intelligence (AI), remote sensing, and genomic surveillance, into conservation practices. These cutting-edge tools significantly enhance our capacity to predict, monitor, and respond to health threats, providing timely insights into disease dynamics and environmental changes.

Innovative technologies like AI can analyze vast datasets to detect patterns, forecast outbreaks, and optimize conservation strategies. Remote sensing offers high-resolution monitoring of ecosystems, enabling researchers to track habitat changes, assess biodiversity health, and identify areas at risk of disruption. Genomic surveillance facilitates the study of pathogens and wildlife populations at the genetic level, improving the detection of zoonotic diseases while informing targeted biodiversity conservation initiatives.

By combining these advancements, *Conservation Medicine* gains the ability to assess ecosystem health in real time, empowering rapid responses to emerging crises. These technologies bridge the gap between scientific research and actionable solutions, ensuring that interventions are both evidence-based and effective. Strategic public-private collaborations ensure the development and deployment of these tools at scale, fostering a global effort to mitigate health threats and protect the interconnected health of humans, animals, and ecosystems.

Moreover, prioritizing equity and inclusion is critical to ensuring that *Conservation Medicine* initiatives are accessible to communities residing in biodiversity hotspots—areas that often host the planet's most vulnerable and marginalized populations. These regions bear the dual burden of being rich in biodiversity yet disproportionately affected by environmental and public health challenges. Supporting capacity-building programs in low- and middle-income countries (LMICs) is imperative to empower local professionals and communities to actively engage in preventing disease outbreaks and safeguarding biodiversity.

By equipping local stakeholders with the necessary tools, resources, and training, these initiatives can foster resilience and self-sufficiency. Integrating local knowledge with scientific expertise enhances the development of culturally sensitive and context-specific solutions that are more likely to be effective and sustainable. Indigenous practices and traditional environmental and ecosystem insights can provide valuable perspectives in designing conservation and public health strategies that align with the unique needs and circumstances of these communities.

Furthermore, targeted investments in education, infrastructure, and research capacity in LMICs can bridge gaps in expertise and access to advanced technologies, enabling equitable participation in global *Conservation Medicine* efforts. A focus on inclusivity not only strengthens the effectiveness of interventions but also ensures that the benefits of conservation and public health initiatives are distributed fairly, fostering long-term success and equity. This approach builds a foundation for collaborative, community-driven solutions that address the interconnected challenges of health and biodiversity conservation.

The challenges to global biodiversity conservation, ecosystem services, and public health systems are daunting, yet they remain surmountable through collective action and determination. A unified global effort, rooted in collaboration, transparent communication, technological advancements, and strategic investments, holds the key to safeguarding the interconnected health of humans, animals, and ecosystems. Though the journey ahead is steep and demanding, the promise of building a sustainable and resilient future through the transformational principles of *Conservation Medicine* is boundless. This moment represents not just an urgent necessity but an unparalleled opportunity to reshape our trajectory and secure the well-being of generations yet to come.

We stand at a critical juncture in Earth's history, where the decisions made today will continue to shape the destiny of our planet. Now is the time for bold and decisive action to ensure a sustainable and thriving world for future generations. We cannot pass the burden of repairing the damage caused by past and present generations onto those yet to come. The responsibility is ours, and the time to act is now!

www.ingramcontent.com/pod-product-compliance
Lightning Source LLC
Chambersburg PA
CBHW071719120626
46550CB00001B/300